LABORATORY MANUAL

CLARINGTON'S HUMAN ANATOMY & PHYSIOLOGY (II)

[FOR ONLINE STUDENTS]

PART TWO

FIRST EDITION

DR. MORRIS B. CLARINGTON

Published by

Conceptual Integrative Solutions Global, LLC
"Gearing Minds for Tomorrow's World"

NOTE FROM AUTHOR

This laboratory manual has been assembled to provide students with an understanding to Biology from a Human Anatomy and Physiology perspective. Students often fell to see the connection between a strong foundation in the Natural Sciences and their success in prospective Healthcare Occupational courses. My goal was to bridge the gap between the Natural Sciences and Healthcare Occupational courses.

Your journey as a healthcare provider begins with you acquiring the necessary skills and knowledge to be proficient in your specialty. Your dedication, compassion, knowledge, attention to detail, and service will one day provide comfort to patients, and offer your life purpose. It is my belief that there is no greater sense of accomplishment than committing one's self to the service and well-being of others.

© 2017 by Dr. Morris B. Clarington. All rights reserved.

Terms of Use Policy

Labs in this manual were compiled from various sources which include free animated, interactive websites on anatomy and physiology, free websites by educators, schools, governments, & nonprofits, and free information offered by commercial websites. Most of the resources are public domain resources and open educational resources. Public domain is a designation for content that is not protected by any copyright law or other restriction and may be freely copied, shared, altered and republished by anyone. Open educational resources (OER) are freely accessible, openly licensed text, media, and other digital assets that are useful for teaching, learning, and assessing as well as for research purposes.

The open educational resources (OER) website hyperlinks to other internet sites featured in this lab manual are listed for your convenience. These other sites are maintained by third parties over which Conceptual Integrative Solutions Global nor your learning institution (school, college, university, etc.) exercises any control. Once you enter these website hyperlinks, whether through an advertisement, another service or content link or through a new browser window, be aware that Conceptual Integrative Solutions Global nor your learning institution is responsible for the privacy practices of these other sites.

The Open educational website hyperlinks may feature materials, information, products, and services provided by third parties. This inclusion of advertisements on these websites does not imply Conceptual Integrative Solutions Global nor your learning institution's endorsement of the advertised products or services. Conceptual Integrative Solutions Global nor the learning institution shall be responsible for any loss or damage of any kind incurred as a result of the presence of such advertisements on these websites. You shall be solely responsible for any correspondence or transactions you have with any third party advertisers.

Every effort has been made to provide you with longstanding, safe, creditable, and reputable, website hyperlinks. If you find that a particular website hyperlink does not work, please notify your instructor so that they may provide you with an alternative website hyperlink that appropriately reflect the concept being taught. We encourage you to review the privacy statements of each website that you visit. If you decide to access any of the third party sites linked, you do so entirely at your own risk.

Some of the materials in this lab manual are original exercises from Conceptual Integrative Solutions Global, and are subject to copyright laws; however, all labs are common to most general biology courses but have been modified to fit the healthcare theme of this manual. No part of this publication may be reproduced, distributed, or transmitted in any form or by any means, including photocopying, recording, or other electronic or mechanical methods, without the prior written permission of the publisher or author. The exception would be in the case of brief quotations embodied in the critical articles or reviews and pages where permission is specifically granted by the publisher or author.

Although every precaution has been taken to verify the accuracy of the information contained herein, the author and publisher assume no responsibility for any errors or omissions. No liability is assumed for damages that may result from the use of information contained within.

Hyperlink policy

This laboratory manual provides hyperlinks to other locations or websites on the Internet. These hyperlinks lead to websites published or operated by third parties who are not affiliated with or in any way related to **Conceptual Integrative Solutions Global or your learning institution (school, college, university, etc.).** They have been included in our manual to enhance your user experience and are presented for information purposes only. We endeavor to select reputable websites and sources of information for your convenience.

However, by providing hyperlinks to an external website or webpage, **Conceptual Integrative Solutions Global nor your learning institution** shall be deemed to endorse, recommend, approve, guarantee or introduce any third parties or the services/ products they provide on their websites, or to have any form of co-operation with such third parties and websites unless otherwise stated.

We are not in any way responsible for the content of any externally linked website or webpage. You use or follow these hyperlinks at your own risk and **Conceptual Integrative Solutions Global nor is your learning institution** responsible for any damages or losses incurred or suffered by you arising out of or in connection with your use of the link. **Conceptual Integrative Solutions Global nor your learning institution** is not a party to any contractual arrangements entered into between you and the provider of the external website unless otherwise expressly specified or agreed to by **Conceptual Integrative Solutions Global or your learning institution**.

Any hyperlinks to websites that contain downloadable software are provided for your convenience only. Again we are not responsible for any difficulties you may encounter in downloading the software or for any consequences from your doing so. Please remember that the use of any software downloaded from the Internet may be governed by a license agreement and your failure to observe the terms of such license agreement may result in an infringement of intellectual property rights of the relevant software provider which we are not in any way responsible.

Please be mindful that when you click on a link and leave our website you will be subject to the terms of use and privacy policies of the other website that you are going to visit.

No Warranties

While every care has been taken in preparing the information and materials contained in this site, such information and materials are provided to you "as is" without warranty of any kind either express or implied. In particular, no warranty regarding non-infringement, security, accuracy, fitness for a particular purpose or freedom from computer virus is given in conjunction with such information and materials.

IMPORTANT: By using this laboratory manual and any of its pages you are agreeing to the terms set out above.

Printed in the United States of America
First Printing, 2017
ISBN-13: 978-1978374164
ISBN-10: 197837416X
For Questions Contact:
Conceptual Integrative Solutions Global, LLC
1-912-495-8745

www.weseesuccess.com

PREFACE

This lab manual was also created to be an affordable option to colleges and/or universities that desire to provide an experimental foundation for the theoretical concepts introduced in the lectures. Labs are intended to reinforce those concepts discussed in lecture.

Many of the labs used are virtual labs. Virtual Labs are web sites or computer software for interactive learning based on simulation of real phenomena. It allows students to explore topics by comparing and contrasting different scenarios, to pause and restart application for reflection and note taking, to get practical experimentation experience over the Internet. When compared to traditional laboratories, virtual laboratories are useful when some experiments involve expensive equipment or when chemicals may cause harm to human health.

For those exercises that are not in virtual space, and require you to purchase inexpensive items, most of the materials used in the experimentation process for this manual can be purchased from most local variety stores.

Labs in this manual are divided into several exercises. This allows an instructor to select those exercises that will best meet their needs. The laboratory manual is designed for students with minimal backgrounds in the physical and biological sciences who are pursuing careers in the healthcare profession. Labs will encompass: Laboratory Safety, Endocrine System, Cardiovascular System, Lymphatic & Immune Systems, Respiratory System, Digestive System, Urinary System, and Reproductive Systems.

Each lab is arranged in three sections: **CRASHCOURSE VIDEO(S)**, **DEFINING KEY TERMS**, and **LAB EXERCISES**.

- **CRASHCOURSE VIDEO(S)** is an educational YouTube channel. The host provides a fast, extensive overview of important anatomical and physiological concepts in a brief 10 to 20 minute video. (Please watch the video link.) **https://youtu.be/hIowxzmCDpw**

- **DEFINING KEY TERMS** refer to important words or expressions that are essential to the student's understanding of the course material or study concept. A definition is a statement expressing the essential nature of something. Definitions enable us to have a common understanding of a word or topic; they allow us to all be on the same page when discussing or reading about a concept. Each word or expression and its definition should be reviewed repeatedly to reinforce rote learning (a memorization technique based on repetition).

- **LAB EXERCISES** are designed to provide students with an experimental foundation for the theoretical concepts introduced in the lectures.

The Open educational resources (OER) used for this book can be collaborated with any of the major online/ distance learning platforms such as: Blackboard, Moodle, or Desire2Learn (D2L).
Please make sure that your computer software is up-to-date.

Computer Requirements:
- **Laptop computer with a minimum of a 2GHz processor and 2GB of RAM**
- **Desktop computer with a minimum of a 2GHz processor and 4GB of RAM**
- **A DSL, cable connection, or greater. (Dial-up is not sufficient)**
- **Windows 7 or newer system operating software for PC computers**

- Disable pop-up blockers
- Antivirus
- Disable all third-party toolbars
- Microsoft Office Suite (Word, Excel, PowerPoint)
- Update your web browser. (Try using Mozilla Firefox, Google Chrome, or Safari)
- Java (Latest Java version)
- Adobe Acrobat Reader
- Adobe Flash Player
- RealPlayer
- Windows: Windows Media Player or Mac: QuickTime
- Audio

Please Note: Your instructor may have the web addresses added as embed codes or hyperlinks that have been integrated into your college's preferred online platform. If this is the case, simply click on the icon or web link to begin the exercises. ONCE YOU ARRIVE TO THE DESIRED WEBSITE ADDRESS, PLEASE DO NOT CLICK ON ANY OF THE THIRD PARTY ADVERTISEMENTS. IF YOU DO SO, YOU MAYBE ROUTED AWAY FROM THE DESIRED WEBSITE.

You will be required to purchase the following items to complete the various exercises in this laboratory manual. Most of the items can be purchased from your local dollar store, if you do not have them in your home already.

Please do not purchase the expensive or name-brand versions of any of these items. This will cut down on your material cost.

- **1 small bottle each of : Coca Cola, Pepsi, Dr. Pepper, Sprite, Mountain Dew and Distilled water**
- 1/2 cup of water
- 1/2 spoonful of crushed chalk
- **6 plastic cups**
- **6 tarnished pennies**
- Coffee filter
- **Pack of drinking straws**
- **Pack of markers**
- **Measuring cup**
- **Notepad**
- One blue Sharpie pen
- One purple Sharpie pen
- Red food coloring
- Rubber band
- Scotch Tape
- Smartphone flashlight app *(Free app)*
- Smartphone with flash
- Two clear glass jars
- **Two Sponges**

TABLE OF CONTENTS

LABORATORY SAFETY RULES & PROCEDURES … p. 9

LAB 1 – THE ENDOCRINE SYSTEM … p. 13

LAB 2 – THE CARDIOVASCULAR SYSTEM … p. 34

LAB 3 – THE LYMPHATIC & IMMUNE SYSTEMS … p. 82

LAB 4 – THE RESPIRATORY SYSTEM … p. 102

LAB 5 – THE DIGESTIVE SYSTEM … p. 118

LAB 6 – THE URINARY SYSTEM … p. 142

LAB 7 – THE REPRODUCTIVE SYSTEM … p. 161

LABORATORY SAFETY RULES & PROCEDURES:

All students must read and understand the information in this section with regard to laboratory safety and emergency procedures prior to the first laboratory session. Your personal laboratory safety depends mostly on you. An effort has been made to address situations that may pose a hazard in the laboratory setting, but the information and instructions provided cannot be considered all-inclusive.

Good common sense is needed for safety in a laboratory. With good judgment, the chance of an accident is very small. Nevertheless, healthcare facilities are full of potential hazards that can cause serious injury and or damage to the equipment.

It is expected that each student will work in a responsible manner and exercise common sense and good judgement. If at any time you are not sure how to handle a particular situation, ask your Instructor for advice. Do not touch anything with which you are not completely familiar. It is always better to ask questions than to risk harm to yourself or damage to the equipment.

ALTHOUGH YOUR LABORATORY EXPERIENCE WILL BE CONFINE TO THE ONLINE LEARNING PLATFORM AND YOUR HOME, IT'S STILL IMPORTANT TO KNOW STANDARD LABORATORY SAFETY GUIDELINES AND PROCEDURES.

General Laboratory Safety Guidelines:
1. No eating or drinking in the laboratory at any time.
2. Playing or "horse play" in the laboratory is forbidden.
3. Read all signs and labels carefully.
4. Use personal protective equipment (PPE) when told by your instructor.
5. Keep the work area clear of all materials except those needed for your work.
6. Be familiar with the location and proper use of fire extinguishers, fire blankets, first aid kits, spill kits, etc. in the laboratory.
7. Make sure to read the assigned experiment before you start to work. Pay close attention to any cautions described in the laboratory application.
8. Students are responsible for the proper disposal of used material. Please make sure to place trash, broken glass, sharps, or any contaminated items in the appropriate containers.
9. Keep pathways clear by placing extra items (books, bags, etc.) on the shelves or under the laboratory tables. If under the tables, make sure that these items cannot be stepped on.
10. Application of cosmetics are prohibited in the laboratory.
11. Never do unauthorized experiments.
12. Keep sinks free of paper or any debris that could interfere with drainage.
13. Do not place fingers or objects, such as pencils, labels, tape, etc., in your mouth when working in the laboratory.
14. Clean up your laboratory work area before leaving.
15. Wash hands with soap and water before leaving the lab and before eating.
16. Follow your instructor's guidelines for using all lab equipment.

17. Notify your instructor if you have any allergies or medical condition that may require special precautionary measures while in the laboratory.
18. In the case of spills: Report at once to your instructor. Cover spilled material with a paper towel and soak with disinfectant. Leave for 20 minutes. Then discard the material in the biohazard buckets.

Accidents and Injuries:
1. Learn the location of the fire extinguisher, eyewash station, first aid kit and safety shower.
2. Report all accidents (spill, breakage, etc.) or injuries (cut, burn, etc.) to the instructor immediately, no matter how trivial it may seem. Do not panic!
3. Long hair (chin-length or longer), dangling jewelry, are hazards in the laboratory. Therefore, long hair must be tied back, and dangling jewelry must be secured. Long hair must be tied back to prevent it from catching fire.
4. If you or your laboratory partner is hurt, immediately (and loudly) call the instructor. Please do not panic!
5. Keep your lab space clean and organized to prevent accidents or injuries.
6. Do not lean, hang over or sit on the laboratory tables.
7. Be careful when lifting heavy objects.
8. Never work alone in the laboratory, unless stated otherwise by your instructor.
9. Do not leave an on-going experiment unattended.
10. Maintain unobstructed access to all exits, fire extinguishers, electrical panels, gas shout offs, emergency showers, and eyewash stations.
11. Do not pour any hazardous material down the sink.
12. Non-disposable contaminated materials are often autoclaved and reused.
13. If someone vomits or cuts him or herself you are to supply only "indirect assistance".
14. During outdoor activities be aware of any dangers in the area, venomous animals, poisonous plants, etc.

Clothing:
1. Any time chemicals, heat, or glassware are used, students will wear safety goggles.
2. Wear safety glasses or face shields when working with hazardous materials, equipment, and/or dissecting specimens.
3. Contact lenses may not be worn in the laboratory.
4. Loose or baggy clothing should be secured so they do not get caught in a flame or chemicals.
5. Lab coats or aprons should be worn during laboratory experiments, unless otherwise stated by your instructor.
6. Shorts and sandals should not be worn in the laboratory at any time, unless otherwise stated by your instructor.
7. Wear gloves when using any hazardous or toxic agent.

Chemical Safety:
1. Treat every chemical as if it were dangerous or hazardous.
2. Make sure all chemicals are clearly and currently labeled.
3. Avoid handling chemicals with fingers.
4. If a chemical should splash in your eye(s) or on your skin, immediately flush with running water for at least 20 minutes. Immediately (and loudly) yell out the instructor's name to get his or her attention.
5. Never remove chemicals or other materials from the laboratory, unless otherwise stated by your instructor.
6. Check the label on all chemical bottles twice before removing any of the contents.
7. Never return unused chemicals to their original container. (Try for the correct amount and share any excess.)
8. Use volatile and flammable compounds only in a fume hood.
9. Never allow a solvent to come in contact with your skin. Always use gloves.
10. Do not taste, or smell any chemicals, especially a solvent. Read the label on the solvent bottle to identify its contents.
11. Dispose of chemical waste according to your instructor's directions.

12. Clean up spills immediately.
13. Know the location of the Material Safety Data Sheet (MSDS) in your laboratory

Handling Glassware and Equipment:
1. Never handle broken glass with your bare hands
2. Use a brush and dustpan to clean up broken glass.
3. Examine glassware before each use.
4. Never use chipped, cracked, or dirty glassware. Check your glassware for cracks and chips each time you use it. Cracks could cause the glassware to fail during use and cause serious injury to you or your lab partners.
5. Do not immerse hot glassware in cold water. The glassware may shatter.
6. Report all breakage of glass or equipment to your instructor immediately.
7. If a piece of equipment fails while being used, report it immediately to your instructor. Never try to fix the problem yourself because you could harm yourself and others.
8. Please make sure to place broken glass in designated container(s) for disposal.
9. Please place anatomical models back in their proper storage places once you are finished handling them, after each class period. Make sure that all moveable or detachable parts are place back in their proper positions.
10. Thoroughly clean all instruments that are exposed to human skin or body fluids.
11. Prepared microscope slides sure be cleaned and placed back in their proper slide set when students are finished handling.
12. Clean and cover microscopes, then place them back into their proper storage place. This ensures that it does its job, while extending its usefulness for as long as possible
13. If you do not understand how to use a piece of equipment, ask your instructor.

Dissecting a Specimen:
1. When making an observation, keep at least one foot away from the specimen.
2. Pointed dissection probes, scalpels, razor blades, scissors, and microtome knives must be used with great care, and placed in a safe position when not in use.
3. Conduct dissections in an appropriate physical environment with the proper ventilation, lighting, furniture, and equipment, including hot water and soap for cleanup. Washing hands is standard operating procedure at all times when a laboratory activity is completed. Without proper ventilation, students will be exposed to potential hazardous vapors from preservatives in specimens.
4. When dissecting smaller specimens, seal the bag after removing the specimen, so as to confine the preservative in the specimen bag.
5. Use personal protective equipment (PPE), such as gloves, chemical splash goggles, and aprons, all of which should be available. PPE is necessary and required when working with materials that can put eyes in harm's way; such as preservatives and body fluids.
6. Body parts or scraps of the specimen are not to be disposed of in the sink.
7. Containers designated for the disposal of sharps (scalpel blades, razor blades, needles; dissection pins, etc.) and containers designated for broken glass are present in each laboratory. Never dispose of any sharp object in the regular trash containers.
8. When cutting with a scalpel or other sharp instrument, forceps may be used to help hold the specimen. Never use fingers to hold a part of the specimen while cutting.
9. Scalpels and other sharp instruments are only to be used to make cuts in the specimen, never as a probe or a pointer.
10. Dispose of dissecting pins or other sharp objects in the red sharps containers/bag, **NOT** in the regular trash.
11. Follow the directions of the instructor concerning the proper disposal of preserved specimens after they are finished being used.

12. Never ingest specimen parts
13. When dissecting, you should cut away from yourself.
14. Treat all living things with respect. Avoid causing unnecessary stress to living animals

Heating Substances:
1. Never use open flames in laboratory unless instructed by instructor.
2. Hair, clothing, and hands should be a safe distance from open flames or hot plates at all times.
3. Heated glassware remain very hot for a long time. They should be set aside in a designated place to cool, and picked up with caution. Allow plenty of time for hot apparatus to cool before touching it.
4. Anyone wearing acrylic nails should not be allowed to work with matches, lighted splints, or bunsen burners.
5. Never look into a container that is being heated.
6. Never point a test tube being heated at another student or yourself. Never look into a test tube while you are heating it
7. Do not place hot apparatus directly on the laboratory desk.
8. If you have a small fire in a container, (for instance, a small beaker full of alcohol has caught fire) find something you can use as a lid for the container. When the container is covered, the fire will quickly burn itself out.
9. If you have a small fire which is not in a container, move away from the fire and shout for help. You can use a fire extinguisher to put the fire out. If you ever need to use a fire extinguisher, remember the following (A) pull the pin, (B) aim to the side at first, (C) depress the handle, (D) sweep the spray from side to side across the BASE of the fire (where the fire meets the fuel), not just at the flames. When the fire is out, clean up the area.
10. Wear safety goggles to protect your eyes when heating substances.
11. If there is a large fire, shout for help and leave the area immediately. The fire alarm will probably sound. When it does, evacuate the building.
12. If your clothing is on fire. Please don't run. It will only fan the flames and make the fire worse. Instead, you should **STOP** moving, **DROP** to the ground (lie down), and **ROLL** on the ground to squash out the flames. Yell continuously. Note: If you want to help a person who is in this sort of trouble, don't use a fire extinguisher. You must never use a fire extinguisher on a human being. The chemicals in the extinguisher can be harmful.

THE ENDOCRINE SYSTEM
LAB 1

CRASHCOURSE VIDEO(S):

Click on the video embedded within your online platform or enter the address below into your web browser:
1. https://youtu.be/eWHH9je2zG4
2. https://youtu.be/SCV_m91mN-Q

(Please make sure to watch the video before continuing)

DEFINING KEY TERMS:

1. Adrenocorticotropic Hormone (ACTH):

2. Aldosterone:

3. Antidiuretic Hormone (ADH):

4. Autopsy:

5. Calcitonin:

6. Cortisol:

7. Endocrine Gland:

8. Epinephrine/ Norepinephrine:

9. Estrogen:

10. Euthanasia:

11. Follicle Stimulating Hormone (FSH):

12. Glucagon:

13. Growth Hormone:

14. Hypothalamus:

15. Insulin:

16. Luteinizing Hormone (LH):

17. Melatonin:

18. Negative Feedback Mechanism:

19. Oxytocin (OT):

20. Parathyroid Hormone (PTH):

21. Positive Feedback Mechanism:

22. Progesterone:

23. Prolactin (PRL):

24. Releasing Hormones

25. Testosterone:

26. Thymosin:

27. Thyroid Stimulating Hormone (TSH):

28. Triiodothyronine (T3) and Thyroxine (T4):

EXERCISE 1.1
ENDOCRINE GLANDS

Purpose of exercise: To provide a general overview of the endocrine system.

Click on *Exercise 1.1* within your online platform or enter the address below into your web browser.
https://medlineplus.gov/ency/anatomyvideos/000048.htm

Please read the overview and watch the video on this website. Use the space below to write a brief summary of what you learned.

_____
_____
_____
_____
_____
_____
_____
_____
_____
_____

_____
_____
_____
_____

EXERCISE 1.2
ANATOMICAL PARTS

Purpose of exercise: To examine the anatomy of the endocrine system.

Click on *Exercise 1.2* within your online platform or enter the address below into your web browser: https://www.biodigital.com/

- *(This is a free site, but you will need to sign up with your name and a validated email address. You can use your personal email address or school email address. MAKE SURE TO CREATE A PASSWORD YOU CAN REMEMBER. I WOULD SUGGEST WRITING IT DOWN AND KEEPING IT IN A SAFE PLACE. YOU WILL USE IT AGAIN.)*

Click **SIGN UP**. Once you have provided the appropriate information, you are now able to access BioDigital's website content. Click **LOG IN** using the email and password you created. Then click **SIGN IN**. On the left of your screen choose from the systems listed.

Please click on the square boxes next to this icon. [icon]. Now click on the **Anatomy By Systems** icon. [icon] Click on the following image(s) with caption. **Male Endocrine System, Female Endocrine System.** View the structures associated with this system.

EXERCISE 1.3
IDENTIFYING ENDOCRINE TISSUES UNDER THE MICROSCOPE

Purpose of exercise: To observe examples of endocrine tissue under the microscope

Click on *Exercise 1.3 (a)* within your online platform or enter the address below into your web browser: http://histologyguide.org/

Once you enter the site, click onto the **Slide Box** tab located on the left hand side. You will see a list of slide categorized by tissue type and organ system in bold font. Click onto the tabs that correctly identifies the tissue type you must observe for this exercise. The tissue type is listed on the table below.

If you encounter difficulties linking to this web address, or would like to view a different source try the link below.

Click on *Exercise 1.3 (b)* within your online platform or enter the address below into your web browser:
http://www.kumc.edu/instruction/medicine/anatomy/histoweb/index.htm

(Your instructor may choose to upload pictures of the tissue types in your college's online platform. If this is the case, you can still utilize the listed slides below as a reference to identify.)

Make a drawing of each prepared slide on high power in the circles below.

Prepared Slide
1. Adrenal
2. Pancreas
3. Parathyroid
4. Pineal
5. Pituitary
6. Thymus
7. Thyroid

Observations

Tissue Type	Tissue Drawing on High Power (40X Objective)
1. Adrenal	
2. Pancreas	

3. Parathyroid	◯
4. Pineal	◯
5. Pituitary	◯
6. Thymus	◯

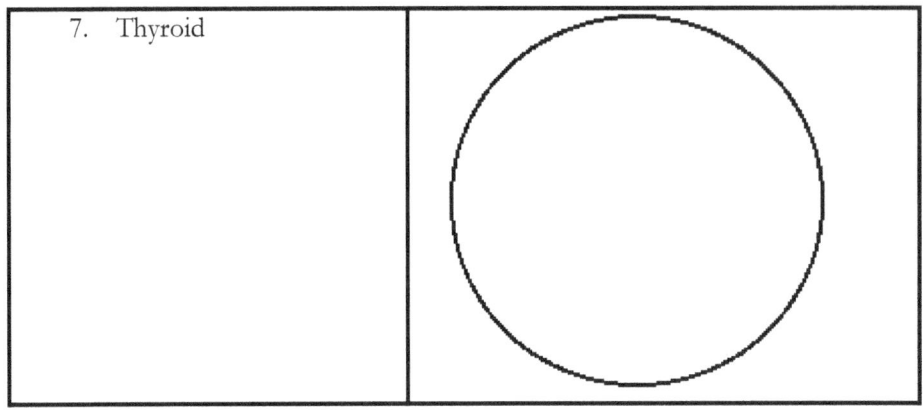

7. Thyroid

EXERCISE 1.4
THE NERVOUS VS. ENDOCRINE SYSTEMS

Purpose of exercise: To contrast the endocrine and nervous systems.

Using your textbook, read about control of the body and negative feedback control.

Complete each statement.

1. Internal control of the body is handled by the _____ system and the _____ system.
2. Most endocrine glands are controlled by the action of the _____, or master gland.
3. A(n) _____ is a chemical released in one part of the body that affects another part.
4. The amount of hormone released by an endocrine gland is determined by the body's _____ for that hormone at a given time.
5. A _____ system is one in which hormones are fed back to inhibit the original signal.
6. When your body is dehydrated, the pituitary releases ADH hormone, which reduces the amount of _____ in your urine.
7. When you have just eaten and your blood glucose levels are high, your pancreas releases the hormone _____, which signals the liver to take in glucose, thereby lowering blood glucose levels.

Comparison	Nervous System	Endocrine System
Fast or slow reaction?		
What system does the signal travel through?		

What is the signal released?		
Which helps regulate the body?		
Which helps with negative feedback?		
Which one is needed for homeostasis?		

EXERCISE 1.5
THE FANTASTICAL WORLD OF HORMONES

Purpose of exercise: To introduce students to the evolution of endocrinology.

Click on *Exercise 1.5* within your online platform or enter the address below into your web browser: **https://topdocumentaryfilms.com/fantastical-world-hormones/**

Please watch the movie. Use the space below to write a brief summary of what you learned. As a leading expert in the study of hormones, John Wass explains the true definition and involvement of the hormones in all of our daily operations. He explains, in more basic terms, the hormone's involvement in our cell stimulation and development. Beyond these personal connections, however, Wass also delves into the importance of the hormone in many scientific breakthroughs and medical treatments, including but not limited to the current, hot topic of obesity and the curing of this epidemic. (If the above video does not show in the hyperlink provided, please type in the name of the video: **The Fantastical World of Hormones with Dr. John Wass** into you google search browser. A query of options for viewing will appear.

_____
_____
_____

EXERCISE 1.6
ACTIONS OF HORMONES

Purpose of exercise: To compare the action of steroids with the action of amino acid-based hormones

Click on *Exercise 1.6* within your online platform or enter the address below into your web browser:
https://www.wisc-online.com/learn/natural-science/life-science/ap13704/the-actions-of-hormones

Read and watch the animation provided on this site.

EXERCISE 1.7
HORMONE MATCHING

Purpose of exercise: To describe the location, hormones, and functions of the following endocrine glands: pituitary, thyroid, parathyroid, adrenal, pancreas, ovaries, testes, pineal, and thymus.

Click on *Exercise 1.7* within your online platform or enter the address below into your web browser.
http://www.zerobio.com/drag_oa/endo.htm

Please read and follow the instructions provided on the website.

EXERCISE 1.8
ENDOCRINE ED

Purpose of exercise: To explain the general mechanisms by which hormones work.

Click on *Exercise 1.8* within your online platform or enter the address below into your web browser:
https://biomanbio.com/HTML5GamesandLabs/Physiogames/endocrine_edhtml5page.html

Please read and follow the instructions provided on the website. After you have used the previous hyperlink move to the next.

EXERCISE 1.9
ENDOCRINOLOGICAL EFFECTS OF PUBERTY

Purpose of exercise: To explain the general mechanisms by which hormones work.

1. At puberty, healthy male gonads (i.e., testicles) will start secreting increased quantities of testosterone. Brainstorm a list of anatomical changes that occur to a young male at puberty? (Or - what is the action of testosterone?)

2. Predict what happens to a young male's <u>behavior</u> during puberty? (For example, how does it affect his interest in sexual activity?)

EXERCISE 1.10
FURTHERING YOUR UNDERSTANDING

Purpose of exercise: To learn about conditions, research, and advancement in endocrinology.

Click on *Exercise 1.10* within your online platform or enter the address below into your web browser:
https://medlineplus.gov/endocrinesystem.html

Click on each endocrine structure to further your understanding. Additional links will be provided as options for viewing. Please click on these hyperlinks and associated with the endocrine structure. Read the articles related to that endocrine structure.

EXERCISE 1.11
ORDERING LABS – FREE T4

Purpose of exercise: To provide students with knowledge about common laboratory tests ordered by healthcare professionals to help diagnose endocrine medical conditions.

Click on *Exercise 1.11* within your online platform or enter the address below into your web browser.
https://labtestsonline.org/understanding/analytes/t4/tab/test/

Answer the questions about the lab test below using the hyperlink/website listed in this exercise. Summarize your answer so that is fits within the space provided.

1. Formal name:
2. Also known as:
3. How is it used?

4. When is it ordered?

5. What does the test result mean?

6. How is the sample collected for testing?

EXERCISE 1.12
ENDOCRINE SYSTEM

Purpose of exercise: To identify endocrine glands and the hormone(s) they secrete.

Directions: Listed below are the major hormones produced by the human body.

ACTH	estrogen	Luteinizing hormone	testosterone
adrenaline	FSH	noradrenaline	thyroxin
aldosterone	glucagons	parathormone	TSH
calcitonin	growth hormone	prolactin	
cortisol	insulin	progesterone	

Next to each gland listed below, write the name of the hormone or hormones it produces.
1. pituitary _____
2. thyroid _____
3. parathyroid _____
4. adrenal _____
5. pancreas (islets of Langerhans) _____

6. testis   _____
7. ovary   _____

Next to each of the functions listed below, write the name of the hormone that produces this effect.

8. raises the blood sugar and increases the heartbeat and breathing rates   _____

9. causes glucose to be removed from the blood and stored   _____

10. influences the development of female secondary sex characteristics   _____

11. promotes the conversion of glycogen to glucose   _____

12. controls the metabolism of calcium   _____

13. promotes the reabsorption of sodium and potassium ions by the kidney   _____

14. influences the development of male secondary sex characteristics   _____

15. stimulates the elongation of the long bones of the body   _____

16. stimulates the secretion of hormones by the cortex of the adrenal glands   _____

17. regulates the rate of metabolism in the body   _____

18. stimulates the development of eggs in the female's ovary   _____

19. involved in the regulation of carbohydrate, protein and fat metabolism   _____

20. stimulates the production of thyroxin   _____

EXERCISE 1.13
ENDOCRINOLOGY VIRTUAL RATS

Purpose of exercise: To provide students with an overview of the endocrine system by working to evaluate autopsy information obtained on virtual rat specimens.

This activity was obtained from the American Physiology Society teaching archive and modified from an activity entitled "Laboratory Exercise Using 'Virtual Rats' to Teach Endocrine Physiology" by Odenweller, CM, et al., Advances in Physiology Education 273:S24-40, 1997.)

1. Using your textbook and what you've learned in class about the regulation of hormones, hormone activity, and how hormone levels might affect the size of a gland, complete the table below.
2. Put a '+' in a cell if you would expect an increase in the size of a gland or in body weight (bottom row); put a '-' in the cell if you would expect to see a decrease in the size of the gland or in body weight if an animal were treated daily with the hormones noted at the top of the table.
3. Leave a cell in the table empty if you do not expect to see a change in a gland's size

	TRH	TSH	ACTH	Cortisol	LH
Pituitary Gland					
Thyroid Gland					
Adrenal glands					
Testes					
Prostate gland					
Body weight					

4. Assume the following scenario to produce the 'virtual rats' whose 'autopsy data' you will analyze: Male rats were either left untreated (control) or were treated daily for two weeks by intravenous injection with one of five hormones.
5. All rats were otherwise treated identically (food, water, etc.). At the end of the two week treatment, all rats were euthanized and subjected to autopsy. The sizes of various endocrine glands was measured and recorded. Overall body weight at the time of euthanasia was also recorded. Unfortunately, due to a miscommunication, all of the rats except the control rat were euthanized together in the same chamber. In other words, at the time of euthanasia, there was no way to determine which of the five experimental rats were treated with which of the five hormones!
6. Use the completed table and the autopsy data provided below for each of the 'virtual rats' in this experiment to help you determine which of the rats was treated with which of the five hormones (TRH, TSH, ACTH, Cortisol or LH).

Except for the control, each rat was treated with only one hormone. Use the values obtained for the control rat as a comparison. To be considered significant, any change observed between the control rat and the experimental rat should be >20% or more. If a change is less than 20%, it is attributed to normal biological differences between individuals or in errors in measurement due to calibration limitations of the measuring equipment.

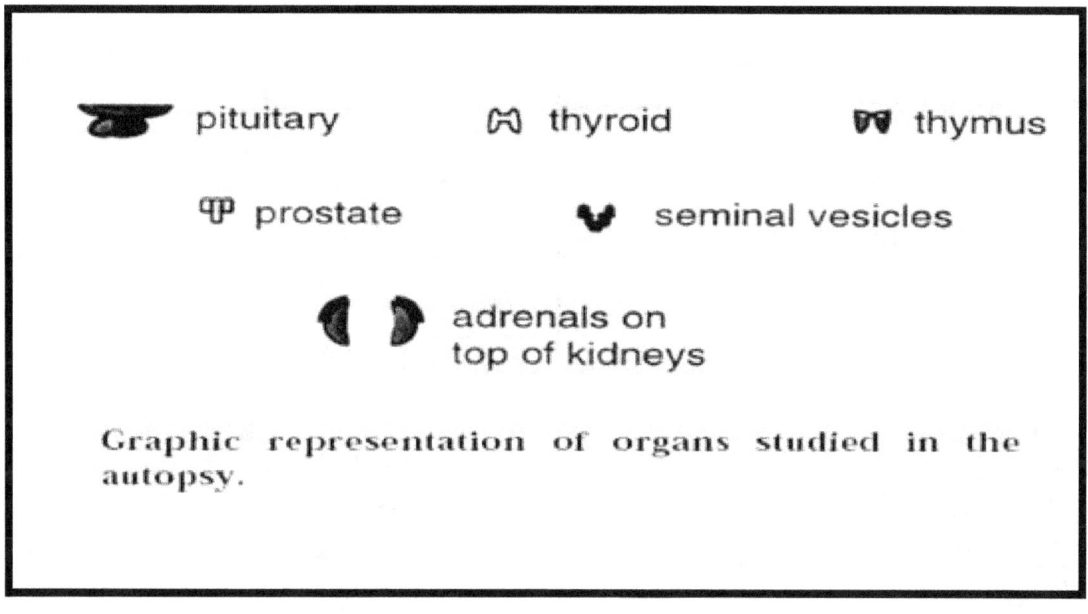

Graphic representation of organs studied in the autopsy.

Control (intact)

Pituitary: 12.9 mg

Thyroid: 250 mg
Thymus: 475 mg

Adrenals: 40 mg
Seminal vesicles: 500 mg
Prostate: 425 mg
Testes: 3200 mg
Body weight: 300 g

Rat 1

Hormone 1 (intact)

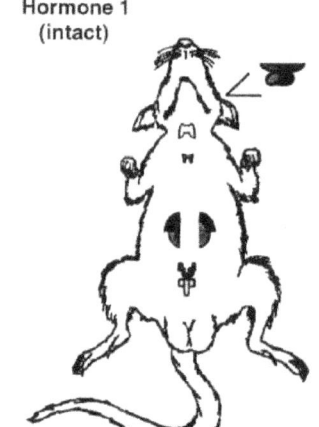

Pituitary: 10.1 mg

Thyroid: 245 mg
Thymus: 250 mg

Adrenals: 100 mg
Seminal vesicles: 490 mg
Prostate: 430 mg
Testes: 3000 mg
Body weight: 200 g

Rat 2

Hormone 2 (intact)

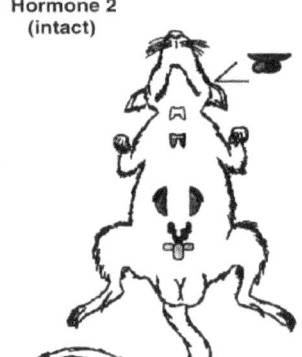

Pituitary: 9.8 mg

Thyroid: 250 mg
Thymus: 480 mg

Adrenals: 40 mg
Seminal vesicles: 900 mg
Prostate: 800 mg
Testes: 5700 mg
Body weight: 385 g

Rat 3

Hormone 4 (intact)

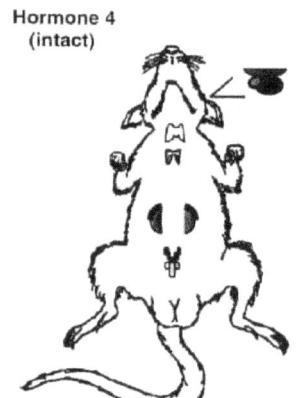

Pituitary: 25 mg

Thyroid: 490 mg
Thymus: 462 mg

Adrenals: 39 mg
Seminal vesicles: 480 mg
Prostate: 400 mg
Testes: 3150 mg
Body weight: 160 g

Rat 4

Hormone 5 (intact)

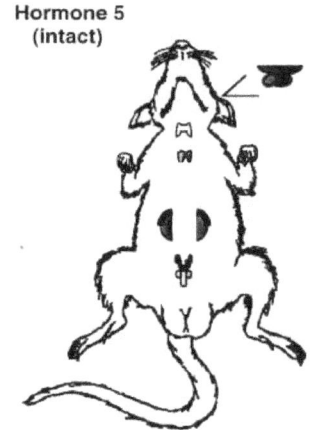

Pituitary: 9.8 mg

Thyroid: 245 mg
Thymus: 150 mg

Adrenals: 30 mg
Seminal vesicles: 475 mg
Prostate: 410 mg
Testes: 3200 mg
Body weight: 150 g

Rat 5

Hormone 6 (intact)

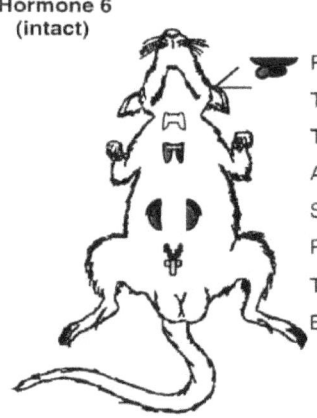

Pituitary: 8 mg
Thyroid: 500 mg
Thymus: 455 mg
Adrenals: 37 mg
Seminal vesicles: 480 mg
Prostate: 405 mg
Testes: 2790 mg
Body weight: 152 g

Rat 1 was most likely treated with: _____

Rat 2 was most likely treated with: _____

Rat 3 was most likely treated with: _____

Rat 4 was most likely treated with: _____

Rat 5 was most likely treated with: _____

EXERCISE 1.14
ENDOCRINE DISEASES AND DISORDERS

Purpose of exercise: To provide students with an understanding of endocrine system disorders.

The following are case histories on each of five difference people. Your job as a potential doctor is to consider the symptoms of each person and then to state in your medical opinion the name of the disorder, the endocrine gland involved, the hormone and whether a hypo- or hyper- secretion is responsible.

1. Laverne had always been pretty, even as a child. Now, even though she is 40 ("over the hill"), she still retained her beauty until last year. Suddenly, she found that her lower jaw seemed to be growing larger. Another thing that she noticed was that her wedding band did not fit anymore. Her hands seemed to be getting bigger! She could not wear the gloves she had worn last year. Also, her nose seemed to be taking over her face! None of her shoes fit her anymore. She used to wear a size 7, now she was up to an 8! Finally she went to her family doctor. If you were her doctor, what would you tell her?

2. Ronnie has always been different from his classmates. When he was in fourth grade, he still could not add or subtract numbers larger than 10. He had trouble reading and spelling the simplest words. His memory was poor. All of the kids constantly teased him and called him names. Finally his parents sent him to a special school. When he got to seventh grade, he had to take gym. After the class all of the boys had to take a shower. Once again the teasing and the tormenting started for poor Ronnie. The other boys told him he looked like a baby and that he would never grow up to be a real man! When he got home, his mother tried to comfort him. Things just continued to get worse. Ronnie kept gaining weight and the nickname "Tubby" was given to him by some cruel classmates. Still another problem was that Ronnie was at least a foot shorter than other boys his age. What was Ronnie's problem?

3. Michael was a pretty normal individual. He was 26 years old, had just gotten married, and had a good job. Everything seemed to be going wonderful when he got hit by some strange disorders. He noticed that he could not bend so easily anymore. In fact, anything that required movement became increasingly painful for him. And then out of nowhere he began to go into muscle spasms. The doctor thought he might have epilepsy, but it was ruled out. His convulsions continued and pretty soon he lost his job and his wife divorced him. What would you have told him about his condition before he lost his job and his wife?

4. Mackenzie was 20 years old. She had a steady boyfriend whom she planned to marry after she was finished with college. She had just been chosen as homecoming queen and was the president of her sorority. Mackenzie was on "cloud nine", nothing could possibly go wrong. Then she noticed her throat had grown larger in a certain area, but she decided that it was due to a bad cold that she had just recovered from. As time went on, however, the swelling continued. First it was the size of a lemon, then a tomato, and now it looked like she had swallowed a grapefruit. People began giving her strange looks, and her boyfriend wasn't too excited about it either. Then her eyeballs seemed to be coming out of her sockets, and her boyfriend began calling her "bug-eyes". What could you have told Mackenzie that would have saved her from all of that embarrassment and heartache?

5. Victoria was a very active 16-year old sophomore. She was involved in clubs and organizations both in and out of school. Whenever possible, she babysat after school to earn some extra money. Three months ago, she started to feel more tired than usual. She had to quit some of her activities and stopped babysitting almost entirely. During the day, Victoria found herself falling asleep in some of her classes and making frequent stops at both the drinking fountain as well as the bathroom. During a urinalysis lab in biology class, her urine tested positive for the presence of glucose. What is Victoria's problem and how should she deal with it?

EXERCISE 1.15
SIGNAL TRANSDUCTION (2 MESSENGER)

Purpose of exercise: To explain the general mechanisms by which hormones work.

Click on *Exercise 1.15* within your online platform or enter the address below into your web browser.
http://www.wiley.com/college/pratt/0471393878/student/animations/signal_transduction/index.html

Please read and follow the instructions provided on the website.

EXERCISE 1.16
FEEDING AND FASTING (ENDOCRINE)

Purpose of exercise: To discuss the functions of the endocrine system in maintaining homeostasis.

Click on *Exercise 1.16* within your online platform or enter the address below into your web browser.
http://learn.genetics.utah.edu/content/metabolism/regulation/

Please read the information provided on the website

EXERCISE 1.17
THE ROLE OF INSULIN

Purpose of exercise: To provide an understanding about the effects of insulin secretion on the human body.

Click on *Exercise 1.17* within your online platform or enter the address below into your web browser.
http://www.mechanismsinmedicine.com/site/view/endocrinology/en-a1/the-role-of-insulin-in-the-human-body

Please read and watch the animation on the website.

EXERCISE 1.18
DIABETES AND INSULIN

Purpose of exercise: To provide an understanding about endocrine disorders of hypo- and hypersecretion.

Click on *Exercise 1.18* within your online platform or enter the address below into your web browser.
https://www.nobelprize.org/educational/medicine/insulin/game/insulin.html

Choose either "*I would like to take care of a dog*" or "*I'm already a caretaker of a dog*". Please read and watch the animation on the website.

EXERCISE 1.19
DIABETES TYPE 2 RISK ASSESSMENT

Purpose of exercise: To provide an understanding about endocrine disorders of hypo- and hypersecretion.

Describe endocrine disorders of hypo- and hypersecretion.

Click on *Exercise 1.19* within your online platform or enter the address below into your web browser.
http://www.diabetes.org/are-you-at-risk/diabetes-risk-test/

Please read the information provided on the website.

EXERCISE 1.20
ENDOCRINE SYSTEM

Purpose of exercise: To further an understanding about the endocrine system and its effects on the human body.

Directions:

1. Explain the difference between endocrine and exocrine glands.
2. What are hormones?

3. Where are they released into from glands?

4. What are the tissues called that are regulated by a specific hormone?

5. How do hormones affect target tissues?

6. Which gland and hormone of the endocrine system

    a. Controls growth

    b. Controls glucose metabolism

    c. Controls female appearance

    d. Controls male characteristics

    e. Helps body in emergencies

    f. Controls rate of food metabolism

7. Hormones carried in the blood stream come in contact with all body tissues, so target cells must have a way to recognize a particular hormone. What are two mechanisms how target cells recognize hormones?

8. What kind of hormones are part of the "two-messenger model?

9. What is the job of the second messenger?

10. Endocrine glands do not secrete hormones at a constant rate. What does the rate of secretion depend on?

11. What are the two ways that a gland gets signals to stop, slow down or speed up the production of hormones?

12. What is the mechanism that alters the activity of a gland?

13. How does negative feedback work?

EXERCISE 1.21
PANCREAS

Purpose of exercise: To evaluate the role of the pancreas as an endocrine organ.

Click on *Exercise 1.21* within your online platform or enter the address below into your web browser.
http://www.wesnorman.com/pancreas.htm

Please read the information provided on the website.

EXERCISE 1.22
THE DUAL ROLES OF THE PANCREAS

Purpose of exercise: To provide an understanding about the dual roles of the pancreas.

Click on *Exercise 1.22* within your online platform or enter the address below into your web browser.
http://www.argosymedical.com/Other/samples/animations/Dual%20Roles%20of%20Pancreas/index.html

Please read and watch the animation on the website.

EXERCISE 1.23
INSIDE DIABETES

Purpose of exercise: To provide an understanding about endocrine disorders of hypo- and hypersecretion.

Click on *Exercise 1.23* within your online platform or enter the address below into your web browser.
http://learn.genetics.utah.edu/content/metabolism/diabetes/

Please read and watch the animation on the website.

EXERCISE 1.24
ENDOCRINE GLANDS

Purpose of exercise: To describe the location, hormones, and functions of the following endocrine glands: pituitary, thyroid, parathyroid, adrenal, pancreas, ovaries, testes, pineal, and thymus.

Directions: Label the following parts on the diagram of the human endocrine system.

a. pineal
b. hypothalamus
c. pituitary
d. thyroid
e. parathyroids
f. thymus
g. adrenal
h. pancreas
i. ovary (female)
j. testis (male)

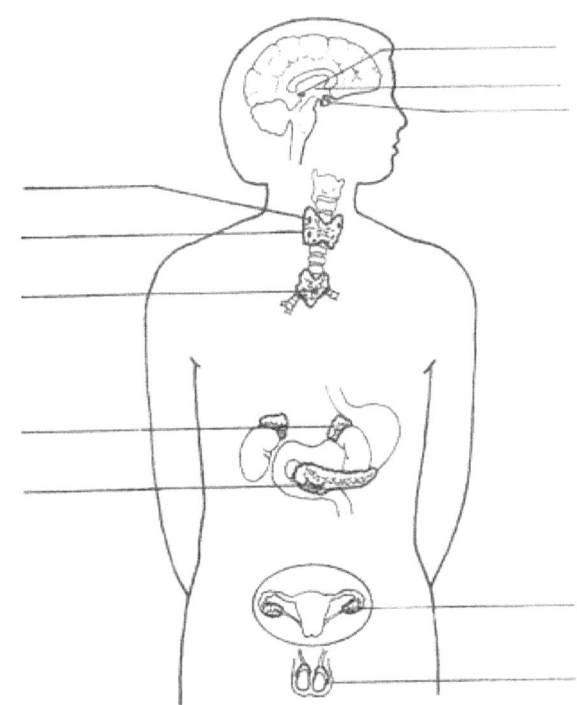

Fill in the blanks with the correct answers.

1. _____ gland may control biorhythms in some animals and control the onset of puberty in humans.

2. _____ gland stimulates metabolic rate and is essential to normal growth and development.

3. _____ gland stimulates growth and stimulates secretion of hormones from other glands.

4. _____ in females stimulates development of secondary sex characteristics, stimulates growth of sex organs at puberty, and prompts monthly preparation of uterus for pregnancy.

5. _____ controls blood glucose levels and determines the fate of glycogen.

6. _____ gland initiates stress responses, increases heart rate, blood pressure, and metabolic rate, dilates blood vessels, mobilizes fat and raises blood sugar levels.

7. _____ gland promotes production and maturation of white blood cells.

8. _____ in males stimulates development of secondary sex characteristics, stimulates growth spurt at puberty, stimulates spermatogenesis.

9. _____ gland increases blood calcium level, stimulates calcium reabsorption and activates vitamin D.

EXERCISE 1.25
WHO AM I?

Purpose of exercise: To identify endocrine glands and the hormone(s) they secrete.

Read each of the following riddles and decide which structure of the endocrine system is being described.

1. _____ I am the vanishing gland. You need me most during your early childhood years and I begin to disappear when you reach puberty. I am considered a member of both the endocrine and the lymphatic system. I secrete a hormone, which helps to stimulate lymphoid cells to produce T-cells. You need me to help fight off diseases. Who am I?
2. _____ I control how "sweet" you are. I keep your blood sugar within normal limits. If you blood sugar is too high I produce insulin and if it is too low, I produce glucagon. I also play a role in the digestion process. Who am I?
3. _____ You can thank me for all those muscles you have and that deep voice. I am also the reason you need to shave every day. I play a role in reproduction by allowing you to make sperms. Who am I?
4. _____ They say "good things come in small packages" and that is true with me. I am very tiny, but I do a lot of jobs in the endocrine system. I help you grow and develop. I also provide the milk for a new mother to breast-feed her baby. The back part of me helps maintain your body's water balance. Finally, when I release my hormone, oxytocin, is will cause the uterus to contract so a new life can be born. Who am I?
5. _____ Many people say I am shaped like a butterfly. I increase metabolism and influence both physical and mental activity. I help with tissue growth. I also cause calcium to be stored in bones. Who am I?
6. _____ There are two of me in your body and I have two parts. I help keep your electrolytes balanced by deciding how much sodium and potassium your body needs. I also play a role in pain control. I am a good friend of the sympathetic nervous system and I play a role in preparing your body to handle emergencies. I help you decide whether to "fight or flight!" Who am I?

THE CARDIOVASCULAR SYSTEM
LAB 2

CRASHCOURSE VIDEO(S):

Click on the video embedded within your online platform or enter the address below into your web browser:
1. https://youtu.be/HQWlcSp9Sls
2. https://youtu.be/9-XoM2144tk
3. https://youtu.be/X9ZZ6tcxArI
4. https://youtu.be/FLBMwcvOaEo
5. https://youtu.be/v43ej5lCeBo
6. https://youtu.be/ZVklPwGALpI

(Please make sure to watch the video before continuing)

DEFINING KEY TERMS:

1. ABO Blood group:

2. Acute Myeloid Leukemia:

3. Agglutination:

4. Agranulocytes:

5. Anastomoses:

6. Anemia:

7. Angioplasty:

8. Antibodies:

9. Anticoagulant:

10. Antigens:

11. Arrhythmias:

12. Atherosclerosis:

13. Atrial syncytium:

14. Basophils:

15. Blood form elements:

16. Cardiac cycle:

17. Cardiomyopathy:

18. Cardiopulmonary resuscitation (CPR):

19. Chronic Myeloid Leukemia:

20. Diabetes Mellitus:

21. Diapedesis:

22. Differential White Blood Cell Count:

23. Dupp sound:

24. Electrocardiogram (ECG or EKG):

25. Embolism:

26. Embolus:

27. Eosinophils:

28. Erythropoiesis:

29. Erythropoietin:

30. Frank-Starling Law of the Heart:

31. Granulocytes

32. Hairy-Cell Leukemia:

33. Hematocrit:

34. Hemocytoblast:

35. Hemostasis:

36. Leukemia:

37. Lubb sound:

38. Lymphocytes:

39. Malaria:

40. Monocytes:

41. Neutrophils:

42. Plasma:

43. Stethoscope:

44. Stroke:

45. Sudden Cardiac Arrest:

46. Thalassemia Major:

47. Thalassemia Minor:

48. Thrombus

49. Ventricular syncytium:

EXERCISE 2.1
CARDIOVASCULAR SYSTEM

Purpose of exercise: To provide a general overview of the cardiovascular system.

Click on *Exercise 2.1* within your online platform or enter the address below into your web browser.
https://medlineplus.gov/ency/anatomyvideos/000023.htm

Please read the overview and watch the video on this website. Use the space below to write a brief summary of what you learned.

_____
_____
_____
_____
_____
_____
_____
_____
_____
_____
_____
_____
_____
_____
_____

EXERCISE 2.2
IDENTIFYING CARDIOVASCULAR TISSUES UNDER THE MICROSCOPE

Purpose of exercise: To observe examples of cardiovascular tissue under the microscope.

Click on *Exercise 2.2 (a)* within your online platform or enter the address below into your web browser:
http://histologyguide.org/

Once you enter the site, click onto the **Slide Box** tab located on the left hand side. You will see a list of slide categorized by tissue type and organ system in bold font. Click onto the tabs that correctly identifies the tissue type you must observe for this exercise. The tissue type is listed on the table below.

If you encounter difficulties linking to this web address, or would like to view a different source try the link below.

Click on *Exercise 2.2 (b)* within your online platform or enter the address below into your web browser:
http://www.kumc.edu/instruction/medicine/anatomy/histoweb/index.htm

(Your instructor may choose to upload pictures of the tissue types in your college's online platform. If this is the case, you can still utilize the listed slides below as a reference to identify.)

Make a drawing of each prepared slide on high power in the circles below.

Prepared Slide
1. Blood Smear
2. Bone Marrow Smear
3. Heart
4. Carotid Artery and Brachiocephalic Vein
5. Vein Valve
6. Aorta and Vena Cava

Observations

Tissue Type	Tissue Drawing on High Power (40X or 60X Objective)
1. Blood Smear	

2. Bone Marrow Smear	
3. Heart	
4. Carotid Artery and Brachiocephalic Vein	
5. Vein Valve	

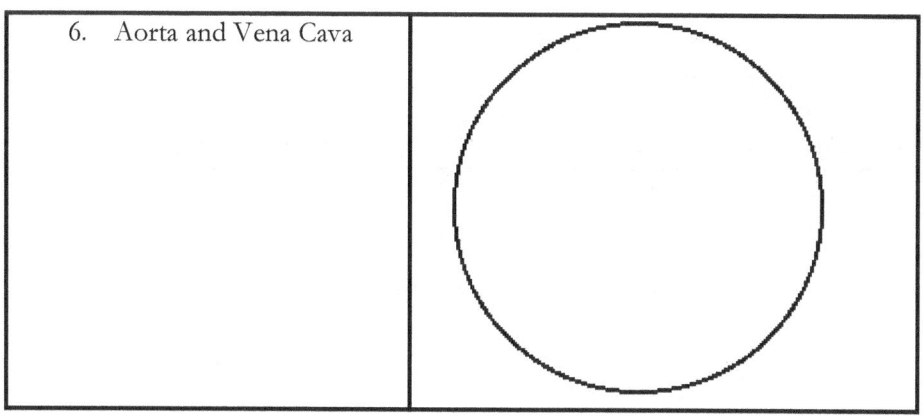

6. Aorta and Vena Cava

EXERCISE 2.3
ANATOMICAL PARTS

Purpose of exercise: To examine the anatomy of the cardiovascular system.

Click on *Exercise 2.3* within your online platform or enter the address below into your web browser: <https://www.biodigital.com/>

- *(This is a free site, but you will need to sign up with your name and a validated email address. You can use your personal email address or school email address. MAKE SURE TO CREATE A PASSWORD YOU CAN REMEMBER. I WOULD SUGGEST WRITING IT DOWN AND KEEPING IT IN A SAFE PLACE. YOU WILL USE IT AGAIN.)*

Click **SIGN UP**. Once you have provided the appropriate information, you are now able to access BioDigital's website content. Click **LOG IN** using the email and password you created. Then click **SIGN IN**. On the left of your screen choose from the systems listed.

Please click on the square boxes next to this icon. [Search Human Library]. Now click on the **Anatomy By Systems** icon. Click on the following image(s) with caption. **Male Cardiovascular System, Female Cardiovascular System.** View the structures associated with this system.

EXERCISE 2.4
OVERVIEW OF THE BLOOD

Purpose of exercise: To provide an overview of the composition of blood.

Click on *Exercise 2.4* within your online platform or enter the address below into your web browser:
https://www.getbodysmart.com/ap/circulatorysystem/blood/menu/menu.html

Choose and view different components of blood by clicking on the **Tutorials** link located on the right of your screen.

EXERCISE 2.5
BLOOD

Purpose of exercise: To illustrate the role of blood in the cardiovascular system.

Click on *Exercise 2.5* within your online platform or enter the address below into your web browser.
http://www.argosymedical.com/Circulatory/samples/animations/The%20Blood/index.html

Please read and watch the animation on the website.

EXERCISE 2.6
RED BLOOD CELL PRODUCTION

Purpose of exercise: To illustrate red blood cell production.

Click on *Exercise 2.6* within your online platform or enter the address below into your web browser.
https://medlineplus.gov/ency/anatomyvideos/000104.htm

Please read the overview and watch the video on this website. Use the space below to write a brief summary of what you learned.

_____
_____
_____
_____
_____
_____
_____
_____
_____

EXERCISE 2.7
HEMATOCRIT TEST

Purpose of exercise: To demonstrate how the hematocrit blood test is used to determine the percentage of red blood cells (RBC's) in blood analysis.

Click on *Exercise 2.7* within your online platform or enter the address below into your web browser.
https://www.getbodysmart.com/ap2/circulatorysystem/blood/rbcs/hematocrit/tutorial.html

Please read and follow the instructions provided on the website. Make sure to click the **blue links** for labels and animations.

EXERCISE 2.8
BLOOD FLOW

Purpose of exercise: To describe blood flow through systemic and pulmonary circuits. Identify the principal arteries and veins of the systemic, pulmonary, and hepatic portal circulations.

Click on *Exercise 2.8* within your online platform or enter the address below into your web browser.
https://medlineplus.gov/ency/anatomyvideos/000012.htm

Please read the overview and watch the video on this website. Use the space below to write a brief summary of what you learned.

EXERCISE 2.9
BLOOD CLOTTING

Purpose of exercise: To illustrate the process of blood clotting.

Click on *Exercise 2.9* within your online platform or enter the address below into your web browser.
https://medlineplus.gov/ency/anatomyvideos/000011.htm

Please read the overview and watch the video on this website. Use the space below to write a brief summary of what you learned.

EXERCISE 2.10
CLOTTING (COAGULATION) TIME TEST

Purpose of exercise: To evaluate the process of blood clotting for clinical analysis. In addition, to identify the stages involved in hemostasis.

Click on *Exercise 2.10* within your online platform or enter the address below into your web browser.
https://www.getbodysmart.com/ap/circulatorysystem/blood/hemostasis/clottingtime/tutorial.html

Please read and follow the instructions provided on the website. Make sure to click the **blue links** for labels and animations.

EXERCISE 2.11
BLOOD TYPING ANIMATION

Purpose of exercise: To illustrate the ABO and Rh blood grouping systems.

Click on *Exercise 2.11* within your online platform or enter the address below into your web browser.
http://waynesword.palomar.edu/aniblood.htm

Please read and watch the animation on the website.

EXERCISE 2.12
BLOOD TYPING GAME

Purpose of exercise: To demonstrate the ABO and Rh blood grouping systems.

Click on *Exercise 2.12* within your online platform or enter the address below into your web browser.
https://www.nobelprize.org/educational/medicine/bloodtypinggame/gamev2/index.html

Please read and follow the instructions provided on the website.

EXERCISE 2.13
ARE YOU MY BLOOD TYPE GAME

Purpose of exercise: To demonstrate the ABO and Rh blood grouping systems.

Click on *Exercise 2.13* within your online platform or enter the address below into your web browser.
http://www.redcrossblood.org/donating-blood/donor-zone/games/blood-type

Please read and follow the instructions provided on the website. Click "**BEGIN**"

EXERCISE 2.14
BLOOD TYPE CALCULATOR

Purpose of exercise: To explain the compatibility of blood types.

Click on *Exercise 2.14* within your online platform or enter the address below into your web browser.
http://www.biology.arizona.edu/human_bio/problem_sets/blood_types/btcalcA_popup.html

Please read and follow the instructions provided on the website. Click on both options.

EXERCISE 2.15
BLOOD TYPE PROBLEMS

Purpose of exercise: To explain the compatibility of blood types.

Use the blood type calculator in Exercise 2.14 to help you answer the questions below.

1. List all the possible genotypes for each of the 4 blood types:
    a. Type O _____
    b. Type A _____
    c. Type B _____
    d. Type AB _____

2. A man with AB blood is married to a woman with AB blood. What blood types will their children be and in what proportion?

3. A man who has type B blood (genotype: BB) is married to a woman with type O blood. What blood type will their children have?

4. A woman with type A blood (genotype: AO) is married to a type B person (genotype: BO). What blood types will their children have?

5. A woman with type A blood is claiming that a man with type AB blood is the father of her child, who is also type AB. Could this man be the father? Show the possible crosses; remember the woman can have AO or AA genotypes.

6. A man with type AB blood is married to a woman with type O blood. They have two natural children, and one adopted child. The children's blood types are: A, B, and O. Which child was adopted?

7. A person with type A blood (unknown genotype) marries a person with type O blood. What blood types are possible among their children. (Show 2 crosses

8. Two people, both with AB blood have four children. What blood types should the children be?

9. A person with type B blood (genotype BO) has children with a type AB person. What blood types are possible among their children?

10. A person with type O blood is married to a person with type A blood (unknown genotype). They have 6 children, 3 of them have type A blood, three of them have type O blood. What is the genotype of the two parents?

11. A person has type B blood. What are ALL the possible blood types of his parents. Show the crosses to prove your answer.

12. A man of unknown genotype has type B blood, his wife has type A blood (also unknown genotype). List ALL the blood types possible for their children. (you may need to do multiple crosses to consider the different possible genotypes of the parents)

13. Two people with type O blood have three children. How many of those three children also have type O blood?

14. Why is a person with type O blood called a "universal donor"?
15. Why is a person with type AB blood called a "universal acceptor"?

EXERCISE 2.16
VIRTUAL BLOOD TYPE

Purpose of exercise: To demonstrate the ABO and Rh blood grouping systems.

Click on *Exercise 2.16* within your online platform or enter the address below into your web browser.
https://www.classzone.com/books/hs/ca/sc/bio_07/virtual_labs/virtualLabs.html

Please read and follow the instructions provided on the website. Record all data on this sheet as you fill them out during the simulation.

Introduction
1. Summarize the problem you are trying to solve.

2. What is the purpose of ths investigation?

Explore Lab
3. What types of serum are available?

   a. What to the antibodies in the serums bind to?

4. What are the disposable micropipettes used for?

Procedure
5. Prediction Chart (indicate YES or NO if you think the blood will clump when exposed to the antibodies)

Blood Type	Anti-A Antibody	Anti-B Antibody	Anti-Rh Antibody
A			
B			
AB			
O			
Rh+			
Rh-			

(Reaction)

Observation Chart

Blood Sample	Clumps in Response to Anti-A	Clumps in Response to Anti-B	Clumps in Response to Anti-Rh	Type
1				
2				
3				
4				

Analyze and Conclude

6. Which, if any, of the blood samples tested cna the patient with type B+ blood receive? Explain why.

7. Explain how you were able to use your knowledge of how different types of blood react with Anti-A, Anti-B, and Anti Rh antibodies to determine the blood types of the four samples.

8. If a person has Type A blood, he or she would have antibodies for what blood type?

9. Why is type O Negative blood known as the universal donor? Why is O Positive not a universal donor?

EXERCISE 2.17
SICKLE CELL ANEMIA VIDEO

Purpose of exercise: To discuss causes of anemia.

Click on *Exercise 2.17* within your online platform or enter the address below into your web browser.
http://www.hhmi.org/biointeractive/sickle-cell-anemia

Please read and follow the instructions provided on the website.

EXERCISE 2.18
HOW RBCS ARE COUNTED (MANUALLY)

Purpose of exercise: To describe selected blood disorders and tests.

Click on *Exercise 2.18 (a)* within your online platform or enter the address below into your web browser.
https://www.getbodysmart.com/ap/circulatorysystem/blood/rbcs/cellnumber/tutorial.html

Click on *Exercise 2.18 (b)* within your online platform or enter the address below into your web browser.
https://www.getbodysmart.com/ap/circulatorysystem/blood/rbcs/totalrbc/tutorial.html

Please read and follow the instructions provided on the website. Make sure to click the **blue links** for labels and animations.

EXERCISE 2.19
WHITE BLOOD CELL DIFFERENTIAL TEST

Purpose of exercise: To describe selected blood disorders and tests.

Click on *Exercise 2.19* within your online platform or enter the address below into your web browser.
https://www.getbodysmart.com/ap/circulatorysystem/blood/wbcs/wbcdifferential/tutorial.html

Please read and follow the instructions provided on the website. Make sure to click the **blue links** for labels and animations.

EXERCISE 2.20
ORDERING LABS – IMMUNOPHENOTYPING

Purpose of exercise: To provide students with knowledge about common laboratory tests ordered by healthcare professionals to help diagnose cardiovascular medical conditions.

Click on *Exercise 2.20* within your online platform or enter the address below into your web browser.
https://labtestsonline.org/understanding/analytes/immunophenotyping/tab/test/

Answer the questions about the lab test below using the hyperlink/website listed in this exercise. Summarize your answer so that is fits within the space provided.

1. Formal name:
2. Also known as:
3. How is it used?

4. When is it ordered?

5. What does the test result mean?

6. How is the sample collected for testing?

EXERCISE 2.21
ORDERING LABS – D-DIMER

Purpose of exercise: To provide students with knowledge about common laboratory tests ordered by healthcare professionals to help diagnose cardiovascular medical conditions.

Click on *Exercise 2.21* within your online platform or enter the address below into your web browser.
https://labtestsonline.org/understanding/analytes/d-dimer/

Answer the questions about the lab test below using the hyperlink/website listed in this exercise. Summarize your answer so that is fits within the space provided.

1. Formal name:
2. Also known as:
3. How is it used?

4. When is it ordered?

5. What does the test result mean?

6. How is the sample collected for testing?

EXERCISE 2.22
IDENTIFYING PATHOLOGICAL BLOOD CONDITIONS UNDER THE MICROSCOPE

Purpose of exercise: To examine blood disorders under the microscope.

Click on *Exercise 2.22* within your online platform or enter the address below into your web browser: http://hematologyatlas.com/principalpage.htm

You will be clicking on the "**ANEMIAS**" and "**LEUKEMIAS**" slides.

(Your instructor may choose to upload pictures of the tissue types in your college's online platform. If this is the case, you can still utilize the listed slides below as a reference to identify.)

Make a drawing of each prepared slide on high power in the circles below. Choose from one of the many slides represented on the website.

Prepared Slide	
1.	Thalassemia Major
2.	Thalassemia Minor
3.	Sickle Cell Anemia
4.	Acute Lymphatic Leukemia
5.	Acute Myeloid Leukemia
6.	Chronic Myeloid Leukemia
7.	Hairy-Cell Leukemia

Observations

Tissue Type	Tissue Drawing on High Power (40X Objective)
1. Thalassemia Major	
2. Thalassemia Minor	
3. Sickle Cell Anemia	
4. Acute Lymphatic Leukemia	

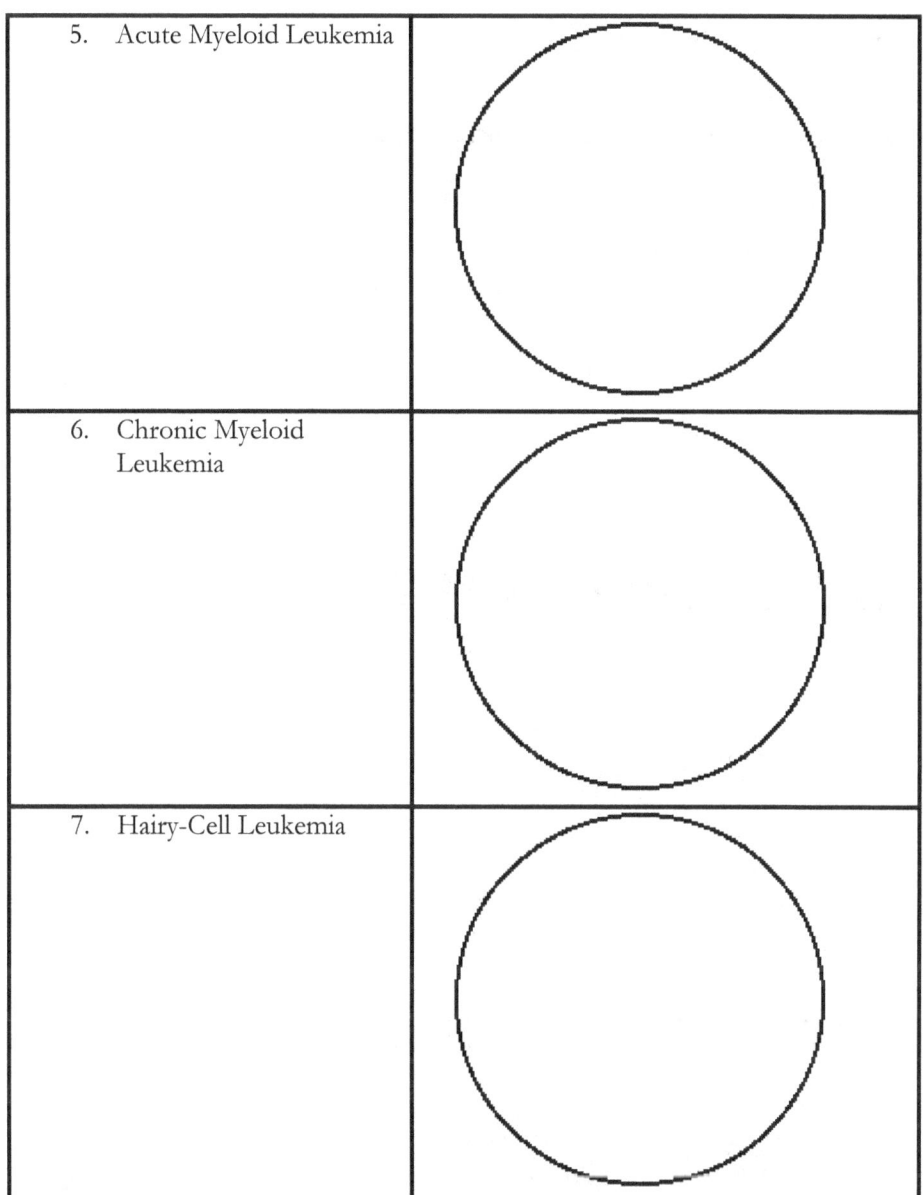

EXERCISE 2.23
EXAMINING BLOOD SMEARS

Purpose of exercise: To study and evaluate blood smears obtained from patients used to diagnose diseases.

Click on *Exercise 2.23* within your online platform or enter the address below into your web browser.
http://glencoe.mheducation.com/sites/dl/free/0078802849/383960/BL_08.html

Please read and follow the instructions provided on the website.

EXERCISE 2.24
PARASITES IN BLOOD

Purpose of exercise: To describe selected blood disorders and tests.

Click on *Exercise 2.24* within your online platform or enter the address below into your web browser.
https://www.nobelprize.org/educational/medicine/malaria/parasite.html

Please read and follow the instructions provided on the website.

EXERCISE 2.25
MALARIA

Purpose of exercise: To describe selected blood disorders and tests.

Click on *Exercise 2.25* within your online platform or enter the address below into your web browser.
https://www.nobelprize.org/educational/medicine/malaria/mosquito.html

Please read and follow the instructions provided on the website.

EXERCISE 2.26
BLOOD PRESSURE

Purpose of exercise: To explain blood pressure and pulse.

Click on *Exercise 2.26* within your online platform or enter the address below into your web browser.
https://medlineplus.gov/ency/anatomyvideos/000013.htm

Please read the overview and watch the video on this website. Use the space below to write a brief summary of what you learned.

_____
_____
_____
_____
_____
_____
_____
_____
_____
_____

_____
_____
_____
_____

EXERCISE 2.27
TAKING BLOOD PRESSURE

Purpose of exercise: To contrast the clinical significance of systolic, diastolic, and pulse pressure.

Click on *Exercise 2.27* within your online platform or enter the address below into your web browser.
https://www.practicalclinicalskills.com/taking-blood-pressure-practice-drill

Please read and follow the instructions provided on the website. Click the "**Click To Take Blood Pressure**" button. Observe systolic and diastolic pressures. Enter your readings and click submit. Take it as many times as you need to master the skill.

You can also use the hyperlink below if you encounter an error with the previous hyperlink.
http://respiratory.guide/bloodPressure/practice-taking-blood-pressure

EXERCISE 2.28
BLOOD PRESSURE CHECKUP

Purpose of exercise: To discuss the factors that affect blood pressure.

Click on *Exercise 2.28* within your online platform or enter the address below into your web browser.
http://www.freemd.com/bp-checkup/overview.htm

Please read and follow the instructions provided on the website.

EXERCISE 2.29
HIGH BLOOD PRESSURE

Purpose of exercise: To explain blood pressure and pulse.

Click on *Exercise 2.29* within your online platform or enter the address below into your web browser.
http://learn.genetics.utah.edu/content/history/bp/

Please read and follow the instructions provided on the website.

EXERCISE 2.30
BLOOD PRESSURE - HYPERTENSION CHECKUP

Purpose of exercise: To discuss the factors that affect blood pressure.

Click on *Exercise 2.30* within your online platform or enter the address below into your web browser.
http://www.freemd.com/hypertension/overview.htm

Please read and follow the instructions provided on the website.

EXERCISE 2.31
BLOOD PRESSURE

Purpose of exercise: To determine what factors affect the likelihood of hypertension.

Click on *Exercise 2.31* within your online platform or enter the address below into your web browser.
http://glencoe.mheducation.com/sites/dl/free/0078802849/383960/BL_08.html

Please read and follow the instructions provided on the website.

EXERCISE 2.32
FINDING YOUR WAY AROUND THE THORAX USING RADIOGRAPHS

Purpose of exercise: To describe the anatomy of the heart and heart wall.

Click on *Exercise 2.32* within your online platform or enter the address below into your web browser.
http://www.wesnorman.com/thoraxradiology.htm

Please read the information provided on the website.

EXERCISE 2.33
OVERVIEW OF THE HEART

Purpose of exercise: To describe the anatomy of the heart and heart wall.

Click on *Exercise 2.33* within your online platform or enter the address below into your web browser.
http://www.wesnorman.com/thoraxlesson4.htm

Please read the information provided on the website.

EXERCISE 2.34
PHASES OF THE CARDIAC CYCLE

Purpose of exercise: To describe the principal events of the cardiac cycle.

Click on *Exercise 2.34* within your online platform or enter the address below into your web browser.
https://www.getbodysmart.com/ap/circulatorysystem/heart/mechanicalevents/cardiaccycle/tutorial.html

Please read and follow the instructions provided on the website. Make sure to click the **blue links** for labels and animations.

EXERCISE 2.35
CARDIAC CONDUCTION SYSTEM

Purpose of exercise: To explain the structural and functional features of the conduction system of the heart.

Click on *Exercise 2.35* within your online platform or enter the address below into your web browser.
https://medlineplus.gov/ency/anatomyvideos/000021.htm

Please read the overview and watch the video on this website. Use the space below to write a brief summary of what you learned.

_____
_____
_____
_____
_____
_____
_____
_____
_____
_____
_____
_____
_____
_____
_____
_____

EXERCISE 2.36
ECG ELECTRODE PLACEMENT: BASIC CONFIGURATION

Purpose of exercise: To explain the structural and functional features of the conduction system of the heart and EKG tracings.

Click on *Exercise 2.36* within your online platform or enter the address below into your web browser.
https://www.getbodysmart.com/ap/circulatorysystem/heart/electricalevents/electrodes/tutorial.html

Please read and follow the instructions provided on the website. Make sure to click the **blue links** for labels and animations.

EXERCISE 2.37
STANDARD BIPOLAR ECG LEAD ELECTRODES

Purpose of exercise: To explain the structural and functional features of the conduction system of the heart and EKG tracings.

Click on *Exercise 2.37* within your online platform or enter the address below into your web browser.
https://www.getbodysmart.com/ap/circulatorysystem/heart/electricalevents/leads/tutorial.html

Please read and follow the instructions provided on the website. Make sure to click the **blue links** for labels and animations.

EXERCISE 2.38
PARTS OF THE CARDIOVASCULAR SYSTEM

Purpose of exercise: To describe the flow of blood through the heart including the pulmonary and systemic circuits.

Label the Diagram
A - Vessels serving the head and upper limbs
B - Vessels serving the body trunk and lower limbs
C - Vessels serving the viscera
D - Pulmonary Circulation
E - Pulmonary "Pump"
F - Systemic "Pump"

Fill in the blanks to trace the path of blood through the circulatory system
From the right atrium to the (1) _____ through the (2)_ ___ valve to the pulmonary trunk to the right and left (3)_____, to the capillary beds of the (4) _____, to the (5) ___, to the (6) _____ of the heart through the (7) ___ valve, to the (8) _____ through the (9) _____ semilunar valve, to the (10) ___, to the systemic arteries, to the (11) ___ of the body tissues, to the system veins, to the (12) _____ and (13) _____, which enter the right atrium of the heart.

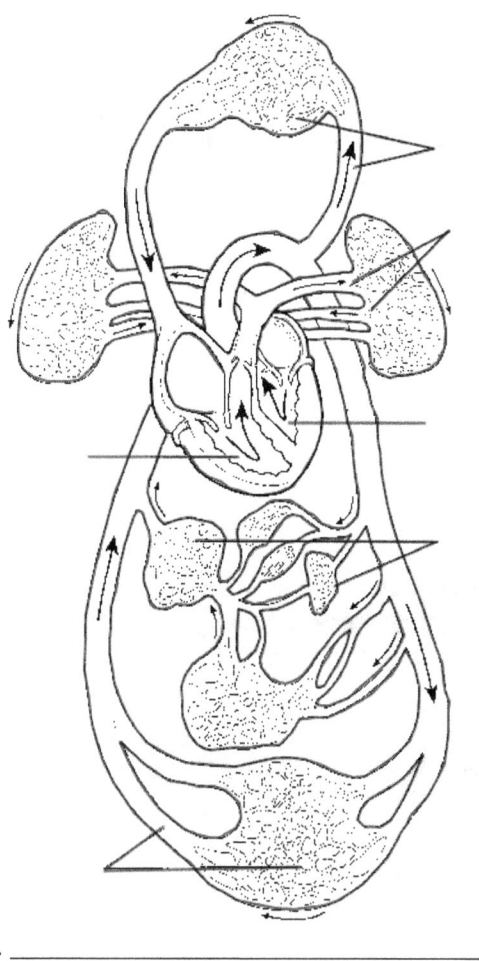

1. _____
2. _____
3. _____
4. _____
5. _____
6. _____
7. _____
8. _____
9. _____
10. _____
11. _____
12. _____
13. _____

EXERCISE 2.39
THE WORKING CARDIOVASCULAR SYSTEM

Purpose of exercise: To describe the flow of blood through the heart including the pulmonary and systemic circuits.

Click on *Exercise 2.39* within your online platform or enter the address below into your web browser.
http://interactivehuman.blogspot.com/2008/10/heart-heart-information-cardiovascular.html

Please read and watch the animation on the website.

EXERCISE 2.40
CARDIOLOGY VIRTUAL LAB

Purpose of exercise: To familiarize you with heritable diseases of the heart. Learn about the diagnostic tools used to examine and diagnose patients.

Click on *Exercise 2.40* within your online platform or enter the address below into your web browser.
http://media.hhmi.org/biointeractive/vlabs/cardiology2/?_ga=2.109733573.1125362674.1506365954-818520368.1502809801

Please read and follow the instructions provided on the website.

EXERCISE 2.41
HEART DISSECTION

Purpose of exercise: To describe the anatomy of the heart and heart wall through dissection.

Click on *Exercise 2.41* within your online platform or enter the address below into your web browser.
http://act.downstate.edu/courseware/haonline/labs/L20/ld0000.htm

Please read the information provided on the website. Click on each step to view the dissection.

EXERCISE 2.42
HEART SURGERY

Purpose of exercise: To identify the risk factors involved in heart disease. In addition, to describe the anatomy of the heart and heart wall.

Click on *Exercise 2.42* within your online platform or enter the address below into your web browser.
http://interactivehuman.blogspot.com/2013/10/heart-surgery-interactive-game-for-kids.html

Please read the information provided on the website.

EXERCISE 2.43
ARRHYTHMIAS

Purpose of exercise: To identify the risk factors involved in heart disease.

Click on *Exercise 2.43* within your online platform or enter the address below into your web browser.
https://medlineplus.gov/ency/anatomyvideos/000005.htm

Please read the overview and watch the video on this website. Use the space below to write a brief summary of what you learned.

_____
_____
_____
_____
_____
_____
_____
_____
_____
_____
_____
_____
_____
_____
_____

EXERCISE 2.44
HEART FAILURE

Purpose of exercise: To describe significant cardiovascular diseases including coronary artery disease and congestive heart failure.

Click on *Exercise 2.44* within your online platform or enter the address below into your web browser.
http://www.heart.org/HEARTORG/Conditions/HeartFailure/Heart-Failure_UCM_002019_SubHomePage.jsp

Answer the questions below using the information from the website.

1. What is heart failure?

2. Name the three types of heart failures.

3. The most common conditions that can lead to heart failure are?

4. What are the signs and symptoms of heart failure?

5. Your treatment plan may include the following.

EXERCISE 2.45
HEART ATTACK

Purpose of exercise: To describe significant cardiovascular diseases including coronary artery disease and congestive heart failure.

Click on *Exercise 2.45* within your online platform or enter the address below into your web browser.
https://youtu.be/idfo87AB2Q8

Please watch the animation on the website.

EXERCISE 2.46
HEART DISEASE

Purpose of exercise: To identify the risk factors involved in heart disease.

Click on *Exercise 2.46* within your online platform or enter the address below into your web browser.
http://learn.genetics.utah.edu/content/history/heart/

Please read and watch the animation on the website.

EXERCISE 2.47
CARDIOMYOPATHY

Purpose of exercise: To describe significant cardiovascular diseases including coronary artery disease and congestive heart failure.

Click on *Exercise 2.47* within your online platform or enter the address below into your web browser.
https://medlineplus.gov/ency/anatomyvideos/000022.htm

Please read the overview and watch the video on this website. Use the space below to write a brief summary of what you learned.

EXERCISE 2.48
CARDIOMYOPATHY - SEPTAL MYECTOMY SURGERY TO TREAT OBSTRUCTIVE HYPERTROPHIC CARDIOMYOPATHY (HCM)

Purpose of exercise: To describe significant cardiovascular diseases including coronary artery disease and congestive heart failure.

Click on *Exercise 2.48* within your online platform or enter the address below into your web browser.
http://nlm.bcst.md/videos/septal-myectomy-surgery-to-treat-obstructive-hypertrophic-cardiomyopathy-hcm?view=displayPageNLM

Please watch the surgical video on this website. Use the space below to write a brief summary of what you learned.

EXERCISE 2.49
CARDIOMYOPATHY - TRANSESOPHAGEAL ECHOCARDIOGRAM (TEE)

Purpose of exercise: To describe significant cardiovascular diseases including coronary artery disease and congestive heart failure.

Click on *Exercise 2.49* within your online platform or enter the address below into your web browser.
http://nlm.bcst.md/videos/transesophageal-echocardiogram-tee?view=displayPageNLM

Please watch the surgical video on this website. Use the space below to write a brief summary of what you learned.

EXERCISE 2.50
SUDDEN CARDIAC ARREST (SCA) RISK ASSESSMENT

Purpose of exercise: To describe significant cardiovascular diseases including coronary artery disease and congestive heart failure.

Click on *Exercise 2.50* within your online platform or enter the address below into your web browser.
https://medlineplus.gov/healthchecktools.html

Please read and follow the instructions provided on the website

EXERCISE 2.51
CARDIOPULMONARY RESUSCITATION (CPR) TRAINING

Purpose of exercise: To describe how to apply the process of CPR during cardiac issues.

Click on *Exercise 2.51* within your online platform or enter the address below into your web browser.
http://interactivehuman.blogspot.com/2012/11/free-medical-training-online.html

Please read and watch the animation on the website.

EXERCISE 2.52
ELECTROCARDIOGRAM (EKG OR ECG)

Purpose of exercise: To explain the structural and functional features of the conduction system of the heart and EKG tracings.

Click on *Exercise 2.52* within your online platform or enter the address below into your web browser.
https://www.nobelprize.org/educational/medicine/ecg/ecg.html

Please read and follow the instructions provided on the website.

EXERCISE 2.53
THE ELECTROCARDIOGRAM (EKG OR ECG)

Purpose of exercise: To explain the structural and functional features of the conduction system of the heart and EKG tracings.

Click on *Exercise 2.53* within your online platform or enter the address below into your web browser.
https://www.getbodysmart.com/ap/circulatorysystem/heart/electricalevents/ecg/tutorial.html

Please read and follow the instructions provided on the website. Make sure to click the **blue links** for labels and animations.

EXERCISE 2.54
READING ELECTROCARDIOGRAM (EKG OR ECG)

Purpose of exercise: To explain the structural and functional features of the conduction system of the heart and EKG tracings.

Please uses the **Cardiac Rhythm Analysis Flow Chart** and the information provided to diagnosis the cardiac condition

1. There are upright smooth and rounded P waves, every P wave is followed by a QRS complex in a 1:1 ratio, the PR interval is < 0.2 seconds in length, and the rate is > 100 bpm. **Name the heart condition?**

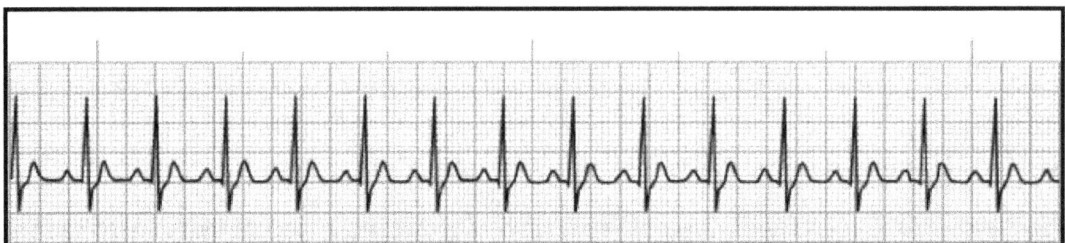

2. There are no clear upright smooth and rounded P waves, there are QRS complexes present and they are all narrow (< 0.1 seconds in width), the rhythm is not regular. **Name the cardiac condition?**

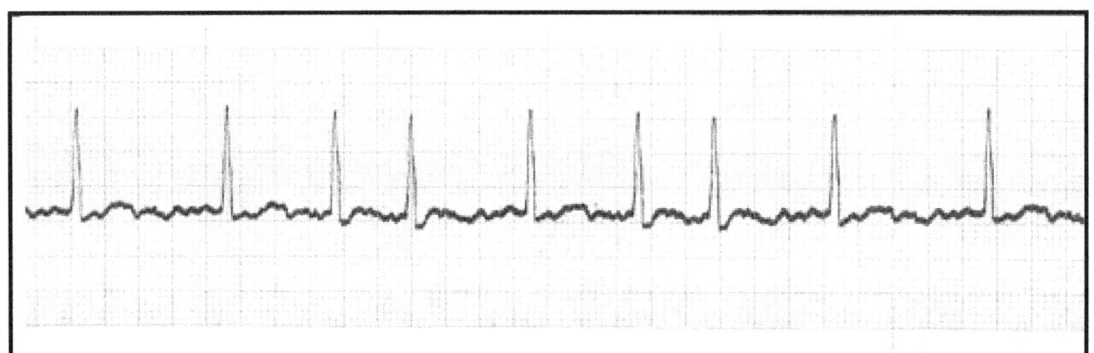

3. There are upright smooth and rounded P waves, there is no 1:1 ratio between the P waves and QRS complex, and the QRS complexes are all broad (> 0.1 seconds in width). **Name the cardiac condition?**

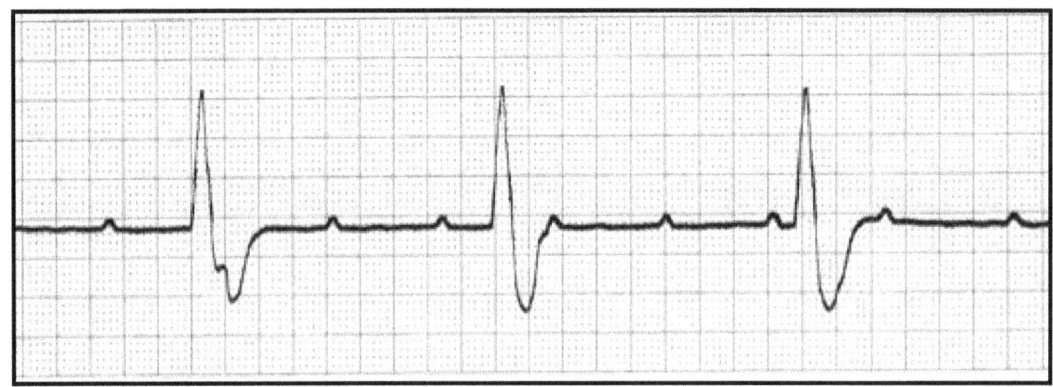

4. There are no clear upright smooth and rounded P waves, there are QRS complexes present and they are all narrow (< 0.1 seconds in width), the rhythm is regular, there are no "saw tooth" flutter waves, the rate is not > 100 bpm, and the rate is between 40 – 60 bpm. **Name the cardiac condition?**

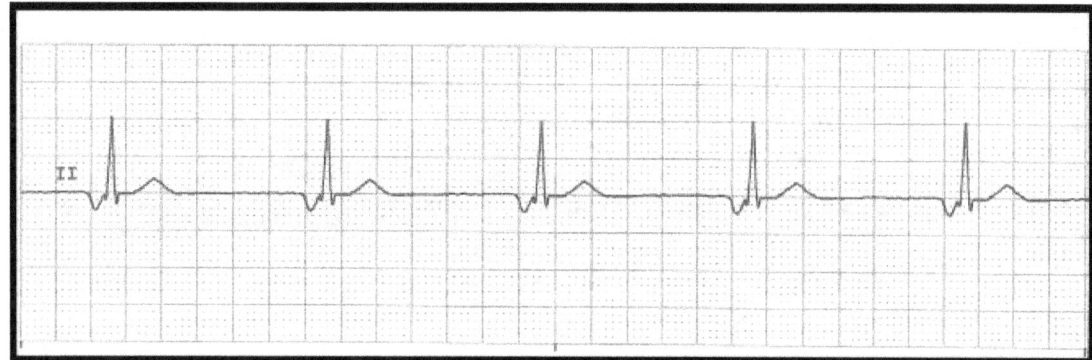

5. There are upright smooth and rounded P waves, not every P wave is followed by a QRS complex in a 1:1 ratio, the QRS complexes are all narrow (< 0.1 seconds in width), and there is a progressive lengthening of the PR interval. **Name the cardiac condition?**

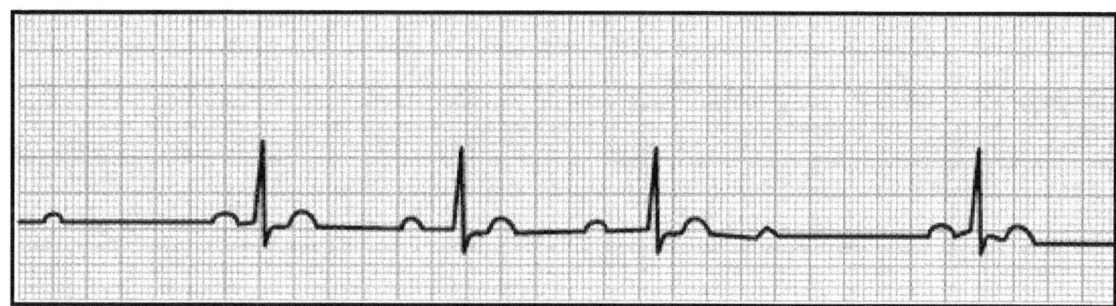

6. There are upright smooth and rounded P waves, not every P wave is followed by a QRS complex in a 1:1 ratio, the QRS complexes are all narrow (< 0.1 seconds in width), and the PR intervals are all the same width. **Name the cardiac condition?**

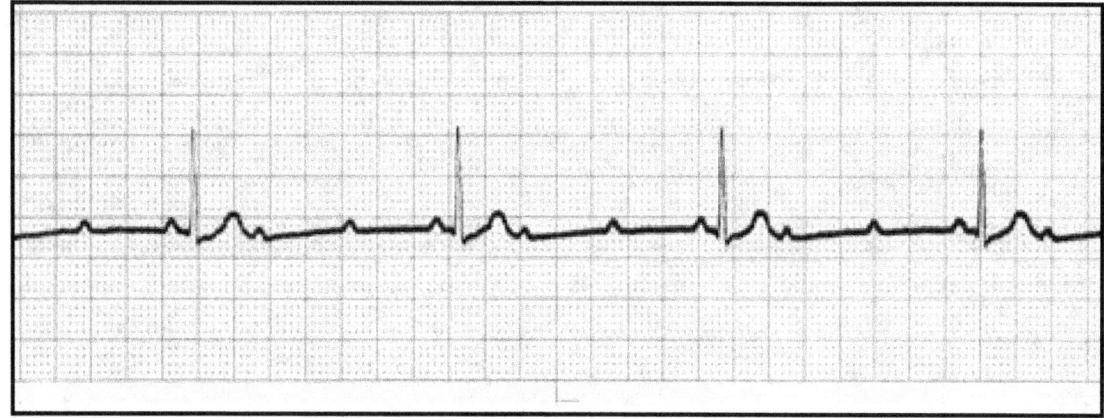

7. There are no clear upright smooth and rounded P waves, there are QRS complexes present and they are all narrow (< 0.1 seconds in width), the rhythm is regular, and there are "saw tooth" flutter waves. **Name the cardiac condition?**

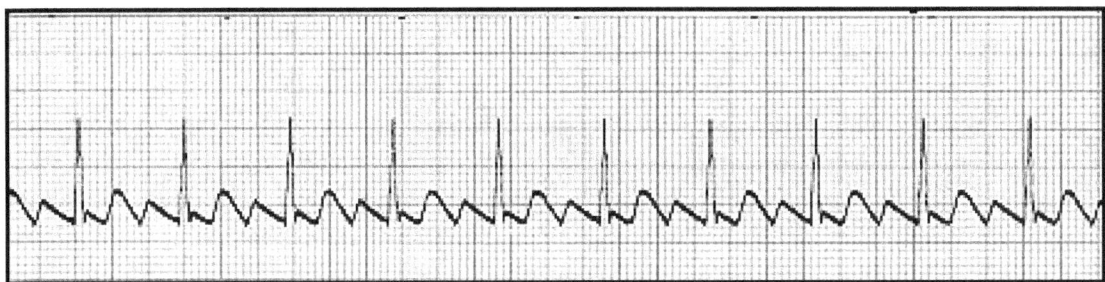

8. There are upright, smooth and rounded P waves, every P wave is followed by a QRS complex in a 1:1 ratio, and the PR interval is > 0.2 seconds in length. **Name the cardiac condition?**

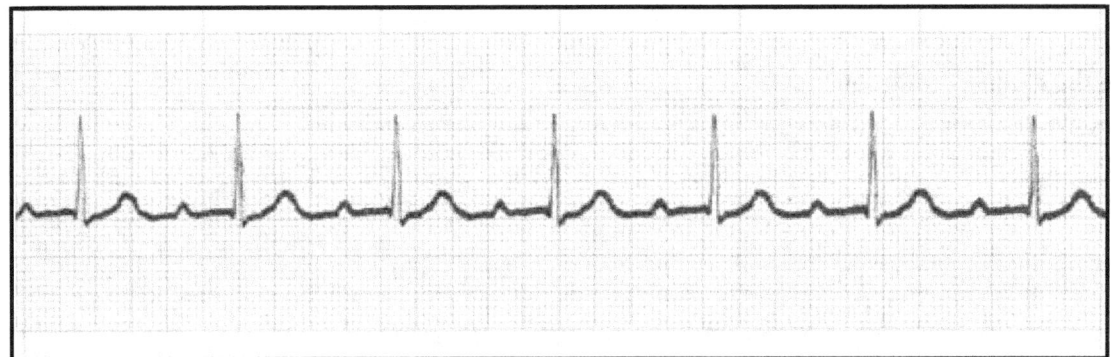

9. There are no clear upright smooth and rounded P waves, there are QRS complexes present and they are all broad (> 0.1 seconds in width), the rhythm is regular, and the rate is < 100 bpm. **Name the cardiac condition?**

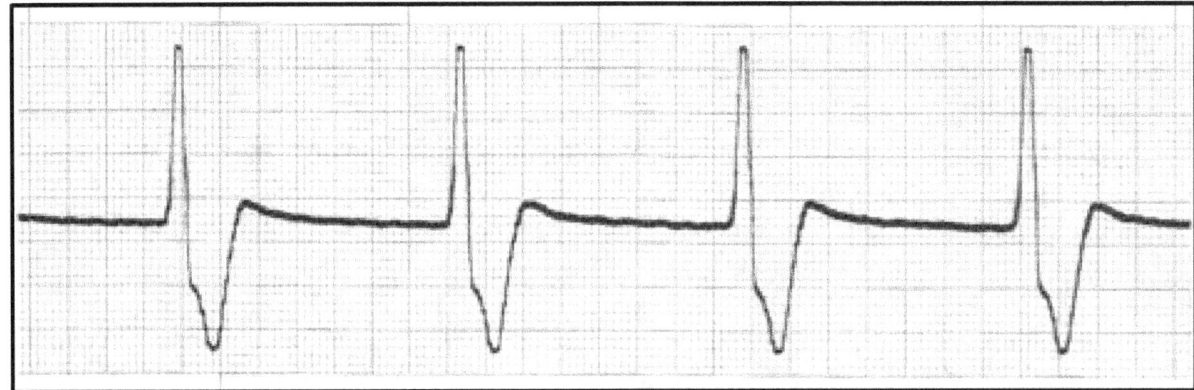

10. There are no clear upright, smooth and rounded P waves, there are QRS complexes present and they are all broad (> 0.1 seconds in width), the rhythm is regular, and the rate is > 100 bpm. **Name the cardiac condition?**

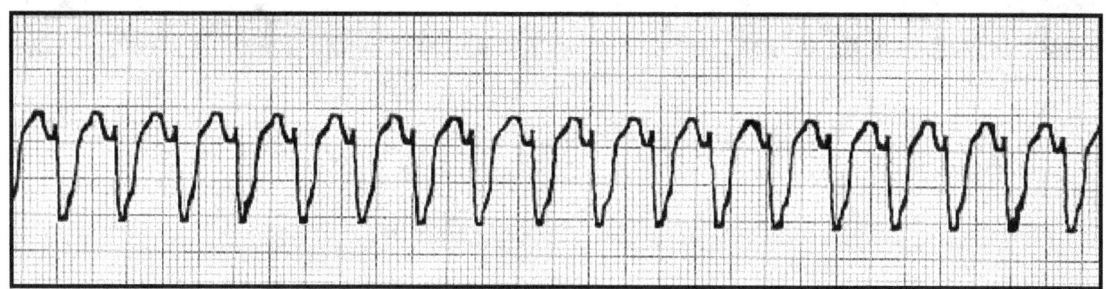

11. There are no clear upright smooth and rounded P waves, there are QRS complexes present and they are all narrow (< 0.1 seconds in width), the rhythm is regular, there are no "saw tooth" flutter waves, and the rate is > 100 bpm, and P waves are indiscernible. **Name the cardiac condition?**

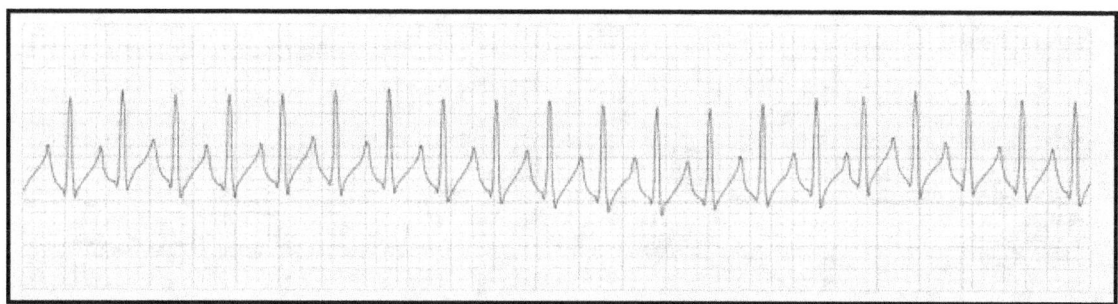

12. There are no clear upright smooth and rounded P waves, there are no clear QRS complexes, and there is electrical activity still evident. **Name the cardiac condition?**

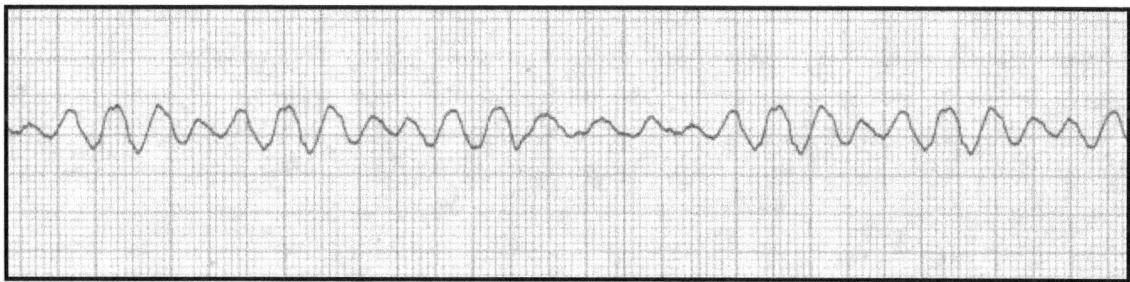

Cardiac Rhythm Analysis Flow Chart

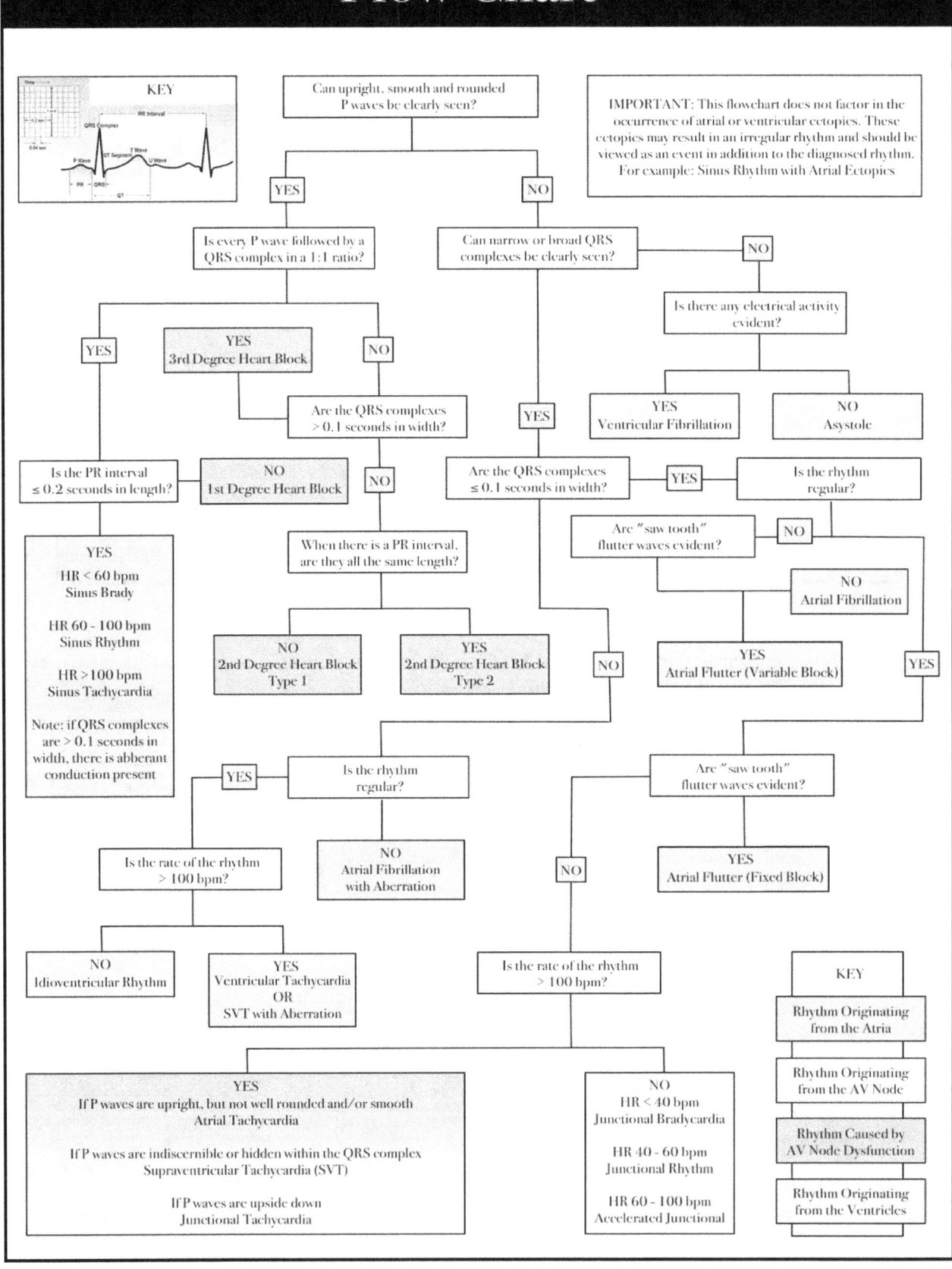

EXERCISE 2.55
ORDERING LABS - TROPONIN

Purpose of exercise: To provide students with knowledge about common laboratory tests ordered by healthcare professionals to help diagnose cardiovascular medical conditions.

Click on *Exercise 2.55* within your online platform or enter the address below into your web browser.
https://labtestsonline.org/understanding/analytes/troponin/tab/test/

Answer the questions about the lab test below using the hyperlink/website listed in this exercise. Summarize your answer so that is fits within the space provided.

1. Formal name:
2. Also known as:
3. How is it used?

4. When is it ordered?

5. What does the test result mean?

6. How is the sample collected for testing?

EXERCISE 2.56
AUTOMATIC IMPLANTABLE CARDIAC DEFIBRILLATOR

Purpose of exercise: To identify medical procedures used to treat cardiovascular disorders.

Click on *Exercise 2.56* within your online platform or enter the address below into your web browser.
http://nlm.bcst.md/videos/automatic-implantable-cardiac-defibrillator?view=displayPageNLM

Please watch the surgical video on this website. Use the space below to write a brief summary of what you learned.

_____
_____
_____
_____
_____
_____

EXERCISE 2.57
HEART BYPASS SURGERY

Purpose of exercise: To identify medical procedures used to treat cardiovascular disorders.

Click on *Exercise 2.57* within your online platform or enter the address below into your web browser.
https://medlineplus.gov/ency/anatomyvideos/000065.htm

Please read the overview and watch the video on this website. Use the space below to write a brief summary of what you learned.

EXERCISE 2.58
IDENTFYING ARTERIES

Purpose of exercise: To contrast the structure and function of the various types of blood vessels.

Click on *Exercise 2.58* within your online platform or enter the address below into your web browser:
http://anatomy.uams.edu/AnatomyHTML/arteries.html

Review the list of vessels provided on the page. Scroll through the list to further your understanding.

EXERCISE 2.59
THE MAJOR SYSTEMIC ARTERIES

Purpose of exercise: To contrast the structure and function of the various types of blood vessels.

Click on *Exercise 2.59* within your online platform or enter the address below into your web browser.
https://www.getbodysmart.com/ap/circulatorysystem/vessels/systemic/majorarteries/tutorial.html

Please read and follow the instructions provided on the website. Make sure to click the **blue links** for labels and animations.

EXERCISE 2.60
IDENTIFYING VEINS

Purpose of exercise: To contrast the structure and function of the various types of blood vessels.

Click on *Exercise 2.60* within your online platform or enter the address below into your web browser:
http://anatomy.uams.edu/AnatomyHTML/veins.html

Review the list of vessels provided on the page. Scroll through the list to further your understanding.

EXERCISE 2.61
THE MAJOR SYSTEMIC VEINS

Purpose of exercise: To identify blood flow through systemic and pulmonary circuits.

Click on *Exercise 2.61* within your online platform or enter the address below into your web browser.
https://www.getbodysmart.com/ap/circulatorysystem/vessels/systemic/majorveins/tutorial.html

Please read and follow the instructions provided on the website. Make sure to click the **blue links** for labels and animations.

EXERCISE 2.62
LABEL THE VESSELS

Purpose of exercise: To identify the principal arteries and veins of the systemic, pulmonary, and hepatic portal circulations.

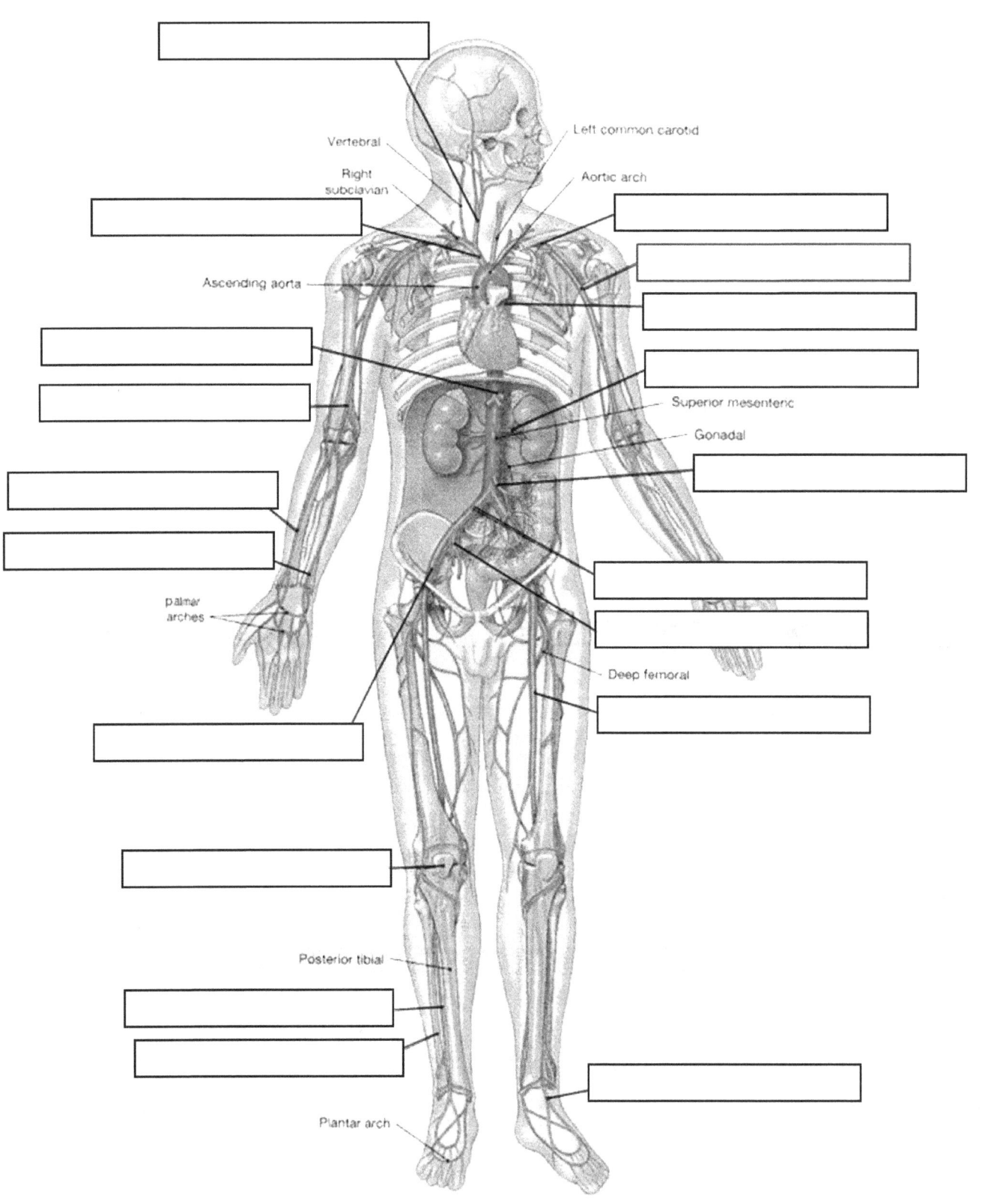

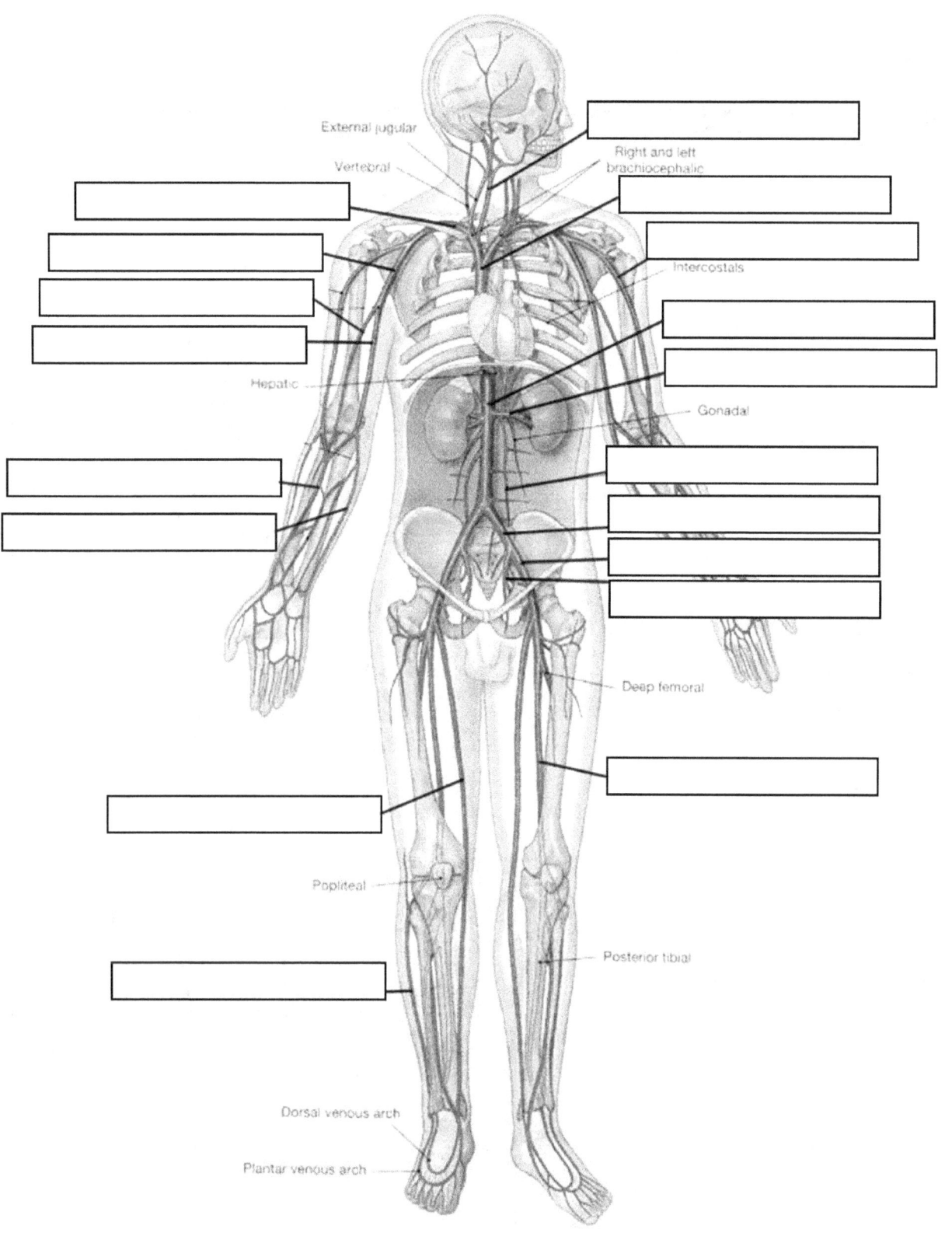

EXERCISE 2.63
IDENTFYING THE MAJOR ARTERIES

Purpose of exercise: To describe blood flow through systemic and pulmonary circuits.

Locate as many arteries as you can on the drawing and color code them by providing their name and colors in the chart below. You should be able to identify at least 20 vessels

1. _____
2. _____
3. _____
4. _____
5. _____
6. _____
7. _____
8. _____
9. _____
10. _____
11. _____
12. _____
13. _____
14. _____
15. _____
16. _____
17. _____
18. _____
19. _____
20. _____

74

EXERCISE 2.64
BLOOD VESSELS AND HEMOSTASIS

Purpose of exercise: To identify the stages involved in hemostasis.

Click on *Exercise 2.64* within your online platform or enter the address below into your web browser.
http://interactivehuman.blogspot.com/2008/05/hemostasis-formation-of-blood-clot.html

Please read and watch the animation on the website.

EXERCISE 2.65
ATHEROSCLEROSIS

Purpose of exercise: To describe significant cardiovascular diseases including coronary artery disease and congestive heart failure.

Click on *Exercise 2.65* within your online platform or enter the address below into your web browser.
https://medlineplus.gov/ency/anatomyvideos/000006.htm

Please read the overview and watch the video on this website. Use the space below to write a brief summary of what you learned.

_____
_____
_____
_____
_____
_____
_____
_____
_____
_____
_____
_____
_____
_____

EXERCISE 2.66
CORONARY ARTERY DISEASE

Purpose of exercise: To describe significant cardiovascular diseases such as coronary artery disease.

Click on *Exercise 2.66* within your online platform or enter the address below into your web browser.
https://medlineplus.gov/ency/anatomyvideos/000037.htm

Please read the overview and watch the video on this website. Use the space below to write a brief summary of what you learned.

_____
_____
_____
_____
_____
_____
_____
_____
_____
_____
_____
_____
_____
_____
_____
_____
_____

EXERCISE 2.67
STROKE

Purpose of exercise: To describe significant cardiovascular diseases.

Click on *Exercise 2.67* within your online platform or enter the address below into your web browser.
https://medlineplus.gov/ency/anatomyvideos/000123.htm

Please read the overview and watch the video on this website. Use the space below to write a brief summary of what you learned.

_____
_____
_____

EXERCISE 2.68
BALLOON ANGIOPLASTY

Purpose of exercise: To identify medical procedures used to treat cardiovascular disorders.

Click on *Exercise 2.68* within your online platform or enter the address below into your web browser.
https://medlineplus.gov/ency/anatomyvideos/000008.htm

Please read the overview and watch the video on this website. Use the space below to write a brief summary of what you learned.

EXERCISE 2.69
PERIPHERAL ARTERY DISEASE (PAD)

Purpose of exercise: To describe significant cardiovascular diseases.

Click on *Exercise 2.69* within your online platform or enter the address below into your web browser.
https://www.heart.org/HEARTORG/Conditions/VascularHealth/PeripheralArteryDisease/Peripheral-Artery-Disease-PAD_UCM_002082_SubHomePage.jsp

Answer the questions below using the information from the website.

1. What is PAD?

2. The most common symptoms of PAD involving the lower extremities are.

3. Is PAD dangerous or life threatening?

4. PAD may be the first warning sign of ?

5. Other symptoms of PAD include..

EXERCISE 2.70
PERIPHERAL ARTERY DISEASE (PAD) - ANGIOPLASTY TO TREAT PERIPHERAL ARTERY DISEASE

Purpose of exercise: To describe significant cardiovascular diseases and means of treatment.

Click on *Exercise 2.70* within your online platform or enter the address below into your web browser.
https://www.broadcastmed.com/4120/videos/angioplasty-to-treat-peripheral-artery-disease-pad?view=displaypageNLM

Please watch the surgical video on this website. Use the space below to write a brief summary of what you learned.

_____
_____
_____
_____

EXERCISE 2.71
THE VIRTUAL AUTOPSY

Purpose of exercise: To teach students how to think critically about information they have learned and how to use it to answer questions about a person's illness or cause of death in case scenarios

Click on *Exercise 2.71* within your online platform or enter the address below into your web browser.
http://www.le.ac.uk/pa/teach/va/titlpag1.html

Please complete the following **Cases: 1, 2, 4, 8, 9, and 17**. Read and follow the instructions provided on the website.

EXERCISE 2.72
REVIEW QUESTIONS

Please continue by answering the questions below.

BLOOD

1. The respiratory gas that is transported mainly in blood plasma is _____.
2. The plasma protein that pulls tissue fluid into capillaries is _____.
3. The gamma globulins in blood plasma are _____ that are produced by the cells called _____.
4. Within RBCs, the essential mineral is _____ because it is part of _____.
5. The hormone _____, which is produced by the _____, stimulates a faster rate than normal of RBC production.
6. The stimulus for the secretion of erythropoietin is _____, and its function is to _____.
7. In RBC formation, iron and protein are necessary nutrients because they become part of _____.
8. Vitamin B_{12} is necessary for RBC formation for the synthesis of _____ in the _____ cells.
9. The life span of RBCs is approximately _____.

10. Bilirubin is made from the _____ of old RBCs, and is excreted by the _____ into _____.
11. The ABO blood types are named for the _____ that are found on a person's _____.
12. The Rh blood group has two major types; these are _____ and _____.
13. A type A person has _____ on the RBCs and _____ in the plasma.
14. A type O person has _____ on the RBCs and _____ in the plasma.
15. A type AB person has _____ on the RBCs and _____ in the plasma.
16. Rh antibodies are formed only when an Rh _____ person is exposed to Rh _____ blood.
17. The kind of WBC that produces antibodies is the _____.
18. The kind of WBC that releases histamine during inflammation is the _____.
19. The kind of WBC that detoxifies foreign proteins is the _____.
20. The kind of WBC that differentiates into a macrophage is the _____.
21. The kind of WBC that recognizes foreign antigens is the _____.
22. The most numerous WBCs are the _____, and the least numerous are the _____.
23. The term _____ means prevention of blood loss.
24. In the process of chemical clotting, the result of the third stage is _____.
25. The vitamin necessary for chemical clotting is _____.
26. The endothelium of a blood vessel is its _____, and it is _____, which prevents abnormal clotting.
27. An abnormal clot that forms within a vessel is called a(n) _____.
28. A clot that breaks off and travels into another vessel is called a(n) _____.
29. The normal pH range of blood is _____ to _____.
30. The term for a low RBC count is _____.

HEART

31. The heart is directly superior to the _____ and is medial to the _____.
32. The outermost of the pericardial membranes is the _____ pericardium.
33. The innermost of the pericardial membranes is the _____ pericardium.
34. The visceral pericardium is also called the _____.
35. The function of the serous fluid of the pericardial membranes is to _____.
36. The lining of the chambers of the heart is the _____.
37. The endocardium is the _____ of the heart, and its function is to _____.
38. The myocardium forms the _____ of the heart.
39. The _____ returns blood from the upper body to the right atrium.
40. The _____ return blood from the lungs to the left atrium.
41. The atria of the heart produce the hormone _____ when blood pressure _____.
42. The _____ emerges from the left ventricle and takes blood to the _____.
43. The _____ emerges from the right ventricle and takes blood to the _____.
44. The aorta takes blood from the _____ ventricle to the _____.
45. The pulmonary artery takes blood from the _____ ventricle to the _____.
46. The tricuspid valve prevents backflow of blood from the _____ to the _____.
47. The aortic semilunar valve prevents backflow of blood from the _____ to the _____.
48. The pulmonary semilunar valve prevents backflow of blood from the _____ to the _____.

49. The coronary arteries are branches of the _____ and supply blood to the _____.
50. The term *systole* means _____ of the heart chambers.
51. The cardiac cycle is the sequence of events in _____.
52. The heart is a double pump: the left side of the heart receives blood from the _____ and pumps it to the _____.
53. The part of the heart that initiates each beat is the _____, which is located in the _____.
54. The normal range of resting heart rate for a healthy adult is _____ to _____ bpm.
55. The electrical impulses for the heartbeat pass from the _____ of the atria to the _____ in the ventricles.
56. The parts of the cardiac conduction pathway in the ventricles, in order, are the _____, _____, and _____.
57. The electrical activity of the heart may be seen in a tracing called a(n) _____.
58. The part of the heart muscle that usually depolarizes first in a heartbeat is the _____.
59. The percent of the blood in a ventricle that is pumped out during systole is called the _____.
60. The part of the brain that regulates heart rate is the _____.

VESSELS

61. The layer of the wall of an artery that is smooth to prevent abnormal clotting is the _____.
62. The layer of the wall of an artery that helps maintain blood pressure is the _____.
63. The layer of the wall of an artery that helps prevent rupture is the _____.
64. Simple squamous epithelium forms the _____ of an artery, and its function is to _____.
65. Smooth muscle tissue forms the _____ layer of an artery, and its function is to _____.
66. Fibrous connective tissue forms the _____ layer of an artery, and its function is to _____.
67. In the vascular system, an alternate pathway for blood flow is provided by vessels called a(n) _____.
68. In capillaries, diffusion is the process by which _____ and _____ are exchanged.
69. In capillaries, nutrients are brought out into tissues by the process of _____.
70. In capillaries, colloid osmotic pressure is created by the presence of _____ in the blood.
71. Arteries carry blood from the _____ to _____.
72. Veins carry blood from _____ to the _____.
73. In pulmonary circulation, blood is pumped to the lungs by the _____, and returns to the _____ of the heart.
74. Veins are able to constrict because of the _____ tissue in their walls.
75. The flow of venous return is kept to one direction only by the _____ in the veins.
76. To compensate for a small loss of blood, the heart rate will _____.
77. A normal blood pressure is considered to be below _____.

THE LYMPHATIC SYSTEM & IMMUNE SYSTEMS
LAB 3

CRASHCOURSE VIDEO(S):

Click on the video embedded within your online platform or enter the address below into your web browser:
1. **https://youtu.be/I7orwMgTQ5I**
2. **https://youtu.be/GIJK3dwCWCw**
3. **https://youtu.be/2DFN4IBZ3rI**
4. **https://youtu.be/rd2cf5hValM**

(Please make sure to watch the video before continuing)

DEFINING KEY TERMS:

1. Adaptive (specific) resistance, or Immunity:

2. Agglutination:

3. Allergic reaction:

4. Antibodies:

5. Antigens:

6. Autoimmune:

7. B cells:

8. Bacteria:

9. Collectins:

10. Complement proteins:

11. Cytokines:

12. Defensins:

13. Fever:

14. Hemotaxis:

15. Immunoglobulins:

16. Inflammation:

17. Innate (nonspecific) resistance:

18. Interferons:

19. Lymph nodes:

20. Lymph:

21. Massage:

22. Natural killer cells:

23. Neutralization:

24. Opsonization:

25. Pathogens:

26. Phagocytosis:

27. Precipitation:

28. T cells:

29. Vaccine:

30. Virus:

EXERCISE 3.1
THE LYMPHATIC SYSTEM

Purpose of exercise: To provide a general overview of the lymphatic system.

Click on *Exercise 3.1* within your online platform or enter the address below into your web browser.
http://www.argosymedical.com/Other/samples/animations/Lymph/index.html

Please read and watch the animation on the website.

EXERCISE 3.2
IDENTIFYING LYMPHATIC TISSUE UNDER THE MICROSCOPE

Purpose of exercise: To observe examples of lymphatic tissue under the microscope.

Click on *Exercise 3.2 (a)* within your online platform or enter the address below into your web browser:
http://histologyguide.org/

Once you enter the site, click onto the **Slide Box** tab located on the left hand side. You will see a list of slide categorized by tissue type and organ system in bold font. Click onto the tabs that correctly identifies the tissue type you must observe for this exercise. The tissue type is listed on the table below.

If you encounter difficulties linking to this web address, or would like to view a different source try the link below.

Click on *Exercise 3.2 (b)* within your online platform or enter the address below into your web browser:
http://www.kumc.edu/instruction/medicine/anatomy/histoweb/index.htm

(Your instructor may choose to upload pictures of the tissue types in your college's online platform. If this is the case, you can still utilize the listed slides below as a reference to identify.)

Make a drawing of each prepared slide on high power in the circles below.

Prepared Slide
1. Lymph Node
2. Palatine Tonsil
3. Spleen
4. Thymus

Observations

Tissue Type	Tissue Drawing on High Power (40X Objective)
1. Lymph Node	

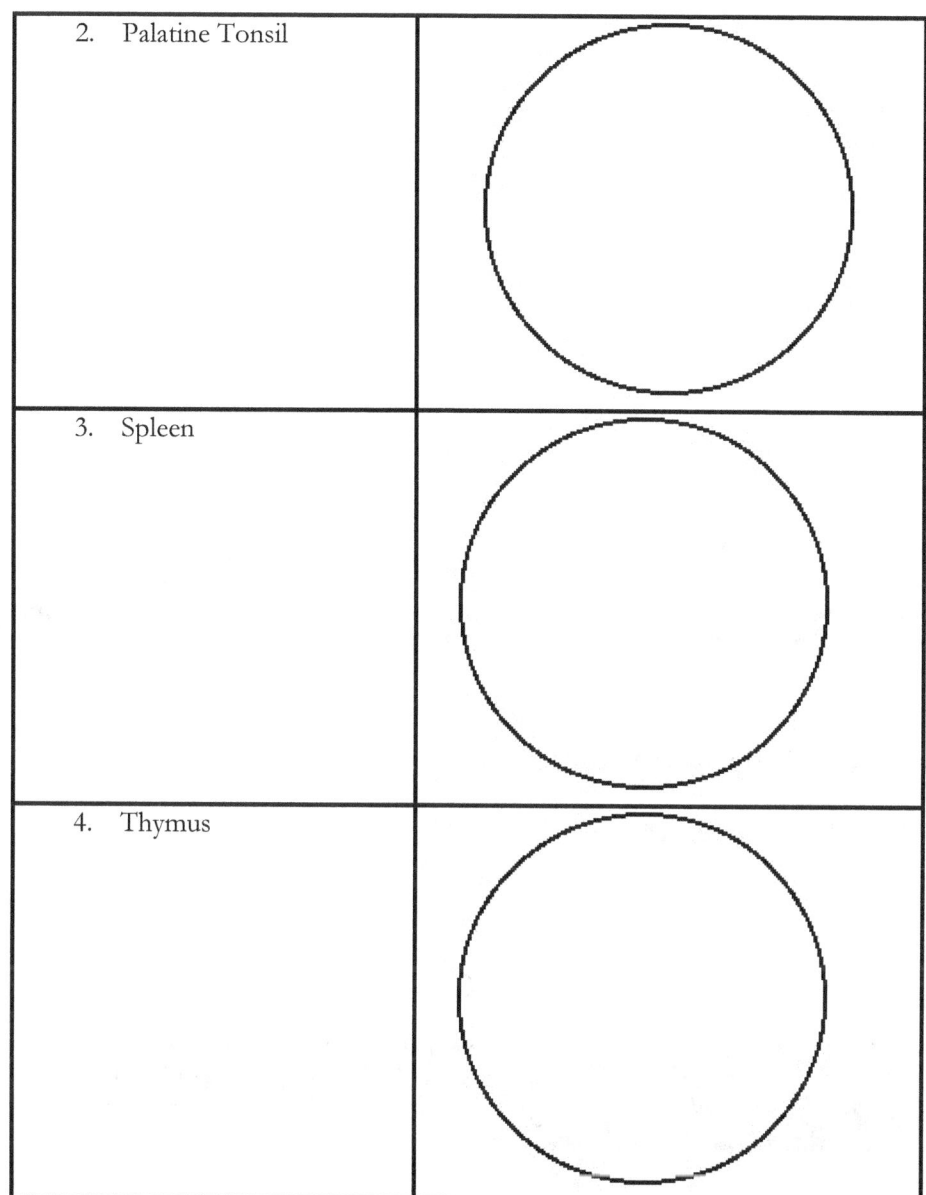

EXERCISE 3.3
ANATOMICAL PARTS

Purpose of exercise: To examine the structures associated with the lymphatic system.

Click on *Exercise 3.3* within your online platform or enter the address below into your web browser:
https://www.biodigital.com/

- *(This is a free site, but you will need to sign up with your name and a validated email address. You can use your personal email address or school email address. MAKE SURE TO CREATE A PASSWORD YOU CAN REMEMBER. I WOULD SUGGEST WRITING IT DOWN AND KEEPING IT IN A SAFE PLACE. YOU WILL USE IT AGAIN.)*

Click **SIGN UP**. Once you have provided the appropriate information, you are now able to access BioDigital's website content. Click **LOG IN** using the email and password you created. Then click **SIGN IN**. On the left of your screen choose from the systems listed.

Please click on the square boxes next to this icon. . Now click on the **Anatomy By Systems** icon. Click on the following image(s) with caption. **Male Lymphatic System, Female Lymphatic System.** View the structures associated with this system.

EXERCISE 3.4
IDENTIFYING LYMPHATIC VESSELS

Purpose of exercise: To identify the names, location, and region of drainage lymphatic capillaries.

Click on *Exercise 3.4* within your online platform or enter the address below into your web browser:
http://anatomy.uams.edu/AnatomyHTML/lymph.html

Review the list of vessels provided on the page. Scroll through the list to further your understanding.

EXERCISE 3.5
SELF-LYMPHATIC DRAINAGE MASSAGE

Purpose of exercise: To demonstrate the lymphatic drainage massage technique. This is a self-massage, which is also known as lymphatic massage, which is used to improve the flow and drainage of lymph by stimulating the lymphatic vessels.

Healthcare providers typically will instruct patients in a program of daily self-massage. This is an important part of managing lymphedema and should be performed regularly as directed. If patients have an infection, or any indication that they are developing an infection, they may have to modify (or skip) their self-massage until the infection is under control. They can return to performing self-massage only after their physician has given clearance.

NOTE: Self-lymphatic drainage massage is **NOT RECOMMENDED** for those with infections or other medical conditions such as a malignant tumor. Get clearance from your healthcare provider before you undertake this exercise.

Introduction:
Your lymphatic system removes fluid build-up and waste from your body and plays an important role in your immune function. It is made up of lymph nodes that are connected by lymph vessels. Large groups or chains of lymph nodes can be found in your neck, under your arms and in your groin Certain health conditions or injuries can prevent fluid from flowing properly, which causes swelling known as lymphedema. Symptoms of lymphedema include swelling in one or more extremities. The swelling may range from mild to severe and disfiguring. While there is no cure for lymphedema, compression treatments, physical therapy, massage therapy can help reduce the swelling and discomfort. By performing lymphatic drainage massage correctly, you can stimulate the opening of the initial lymphatic and increase the volume of

lymph flow by as much as 20 times. Lymphatic drainage massage is a great ally in any therapist's tool kit. By being able to address the lymph system directly, a patient's immune system function can be significantly increased. When we have a strong immune system, we are happy, balanced and whole.

Procedure:
Preparing for the self-lymphatic drainage massage
1. Select a room for your self-massage that is quiet and at a comfortable temperature.
2. Perform your massage in the positions that work best for you. Most commonly this is lying on your back in a comfortable and relaxed position. You may want to remove your glasses, jewelry, and any restrictive clothing.
3. Each self-massage begins with a few quiet moments of deep breathing to help you relax and focus on the task at hand. This breathing pattern should be maintained throughout the massage session.

Understanding the self-lymphatic drainage massage strokes
4. Self-massage is a gentle technique taught to the patient by the therapist and the resulting massage should never hurt or make the skin red. Use a light pressure and keep your hands soft and relaxed. The pressure of your hands on your skin should be just enough to gently stretch the skin as far as it naturally goes, and then releasing. If you can feel your muscles underneath your fingers, then you are pressing too hard.
5. Most self-massage strokes use very little pressure and the hands do not slide over the skin. Instead they move and stretch the skin to stimulate flow of lymph through the lymphatic capillaries that are located just under the skin.
6. Oils and lotions that make the skin slippery are not ordinarily used during self-massage. If the skin is very dry, a lotion can be applied and allowed to absorb before you continue with the massage.
7. Self-massage to encourage lymph drainage is not the same as conventional muscle massage. For this reason, it is important that you do not allow anyone, other than a qualified lymphedema therapist, to massage lymphedema affected tissues using deep strokes.
8. Use the flats of your hands instead of your fingertips. This allows more contact with the skin to stimulate the lymph vessels
9. Try to do the massage when you are comfortably warm because your muscles will be more flexible.
10. Never stimulate the flow of lymph from a normal area into a lymphedema affected area. Massage towards areas of your body that have not been treated for cancer
11. The length of time devoted to self-massage depends on your condition and the instructions provided by your therapist.
12. Once practiced in self-massage, many individuals find that this massage takes only a few minutes.

Preforming the self-lymphatic drainage massage
(NOTE: Steps 1 – 4 stimulate your lymph system and Steps 5-11 address lymphedema symptoms in your legs)

1. **Deep Breathing**
   A very important part of your self-care is deep breathing. Deep breathing helps to stimulate lymphatic system in your whole body. You can practice deep breathing anytime!
   a. Place the palms/flats of your hands on your stomach
   b. Slowly, breathe in deeply through your nose, and let your stomach expand
   c. Breathe out slowly through pursed lips, and let your stomach flatten
   d. Repeat 5 times. Take a short rest between each breath so you do not get dizzy

2. **Stretch and release the skin at the front of your neck**
   This step helps lymph fluid drain back to your bloodstream at your neck. You can massage one side at a time or both sides at the same time. Cross your hands if you are doing both at the same time.
   a. Place the flats of your 2nd and 3rd fingers on either side of your neck, just above your collarbone. Do a shoulder shrug up and feel the dip in the skin. This is the correct position.
   b. Massage down and inwards towards your collarbone. Always keep your fingers above your collarbone. Gently stretch the skin just as far as it naturally goes and release.
   c. This massage will look like two "J" strokes facing one another.
   d. Repeat 15 times.

3. **Stretch and release the skin at the side of your neck**
   You can massage one side at a time, or do both sides together.
   a. Place your flat hands on either side of your neck, just under your ears
   b. Gently stretch the skin back (away from your face) and down, then release.
   c. Try to massage your neck in a slow, gentle way, following a rhythm.
   d. Repeat 10 to 15 times.

4. **Stretch and release the skin on the back of your neck**
   a. Place your flat hands on the back of your neck, just below your hairline on either side of your spine.
   b. Stretch the skin towards your spine and then down towards the base of your neck and release.
   c. Repeat 10-15 times

5. **Prepare your chest**
   This step prepares the lymph nodes in your under arm to take in lymph fluid from your affected abdomen and upper leg. For Step 5 and 6, place your arm in a comfortable position. Keep it slightly raised and supported. Try placing it on an arm rest or table for comfort.
   a. Place your palm against your underarm on the side of your body that has lymphedema.
   b. Gently pull up and in toward your body, then release
   c. Repeat 10 to 15 times.

6. **Stretch and release the skin from your hip to your underarm**
   Do this step on the side of your body that has lymphedema.
   This step will direct fluid away from your hip to your underarm.
   a. Place your hand on your hip on the side that has lymphedema.
   b. Gently massage from the outside of your hip (beside your hip bone) upward along the side of your body to your underarm.
   c. Gently stretch the skin as far as it goes naturally and release.
   d. Repeat 10 to 15 times.

7. **Stretch and release the skin from the inside of your leg to the outside of your leg**
   Fluid normally flows up the inside of your leg into the lymph nodes in your groin. This step directs fluid away from the inside of your leg, up the side of your body into your underarm. When doing this massage, make sure you are comfortable. Do not strain your back, wrists or hands. You can also add deep breathing to make the massage even more effective.
   a. Start at the top of your leg.
   b. Place one hand on the inside of your leg and one hand on the back of your leg, without straining your wrist, hand or arm.
   c. Gently stretch and release the skin from the inside of your leg towards the outside of your leg and up towards your hip.
   d. Shift your hands lower on your leg and repeat. Keep shifting down and repeating this step until you reach your knee as shown in the pictures.
   e. Now place your hand on the outside of your knee. Switching between your left and right hands, stretch the skin in an upward motion towards your underarm.
   f. Repeat each section 10 to 15 times.

8. **Prepare your knee**
   This step will prepare the lymph nodes in the back of your knee to take in fluid.
   a. Place your hands behind your knee
   b. Gently pump the back of your knee in a rolling, upward motion (making a "J" shape).
   c. Repeat 10 to 15 times.

9. **Stretch and release the skin on your lower leg**

    a. Place one hand on your shin and the other hand on the back of your lower leg, just below your knee.
    b. Gently stretch the skin towards your upper leg and release.
    c. Shift your hands down and repeat this upward motion until you reach your ankle.
    d. Remember to stretch and release the skin up towards your knee.
    e. Repeat 10 to 15 times.

10. **Stretch and release the skin on your ankle and foot**
    a. Continue the stroke from the previous step on your ankle and foot.
    b. Remember to gently stretch the skin as far as it goes naturally
    c. Always stroke up.

11. **Massage for swollen toes**
Do the following step if your toes are swollen.
    a. Place your index finger and thumb on the base of your toe.
    b. Gently push the fluid towards your foot.
    c. Repeat often.

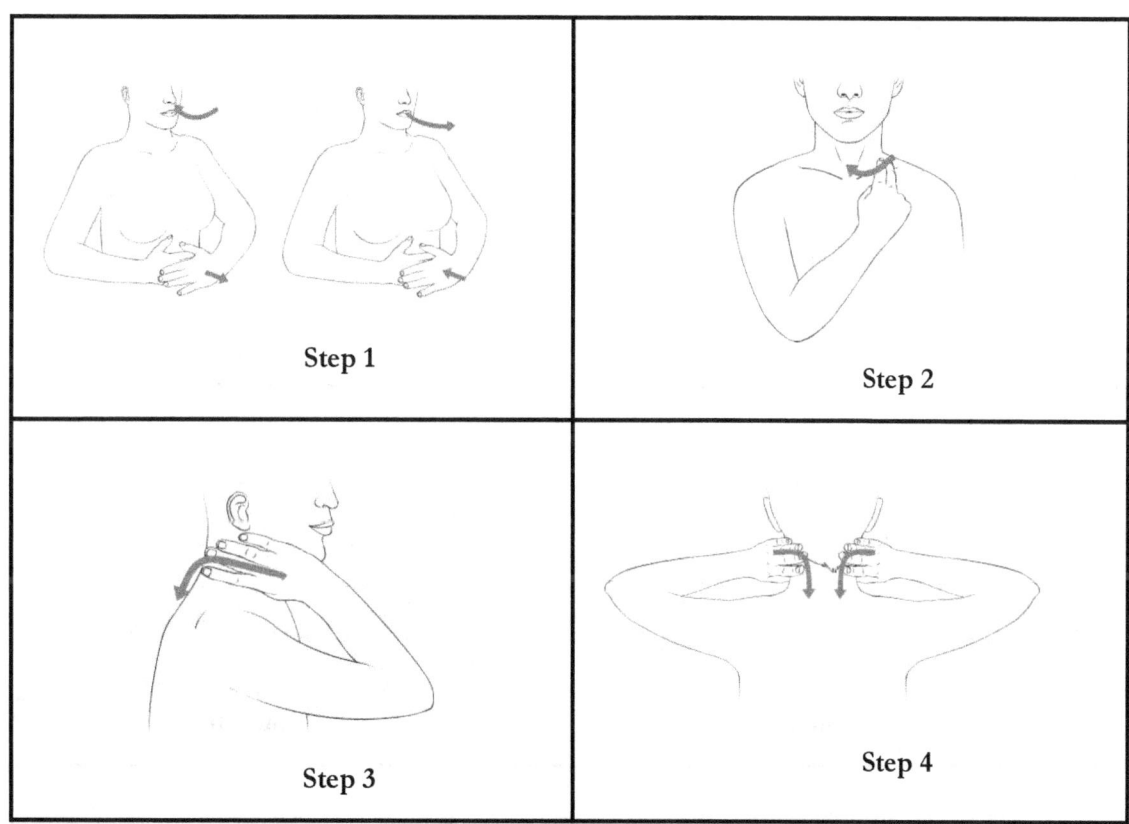

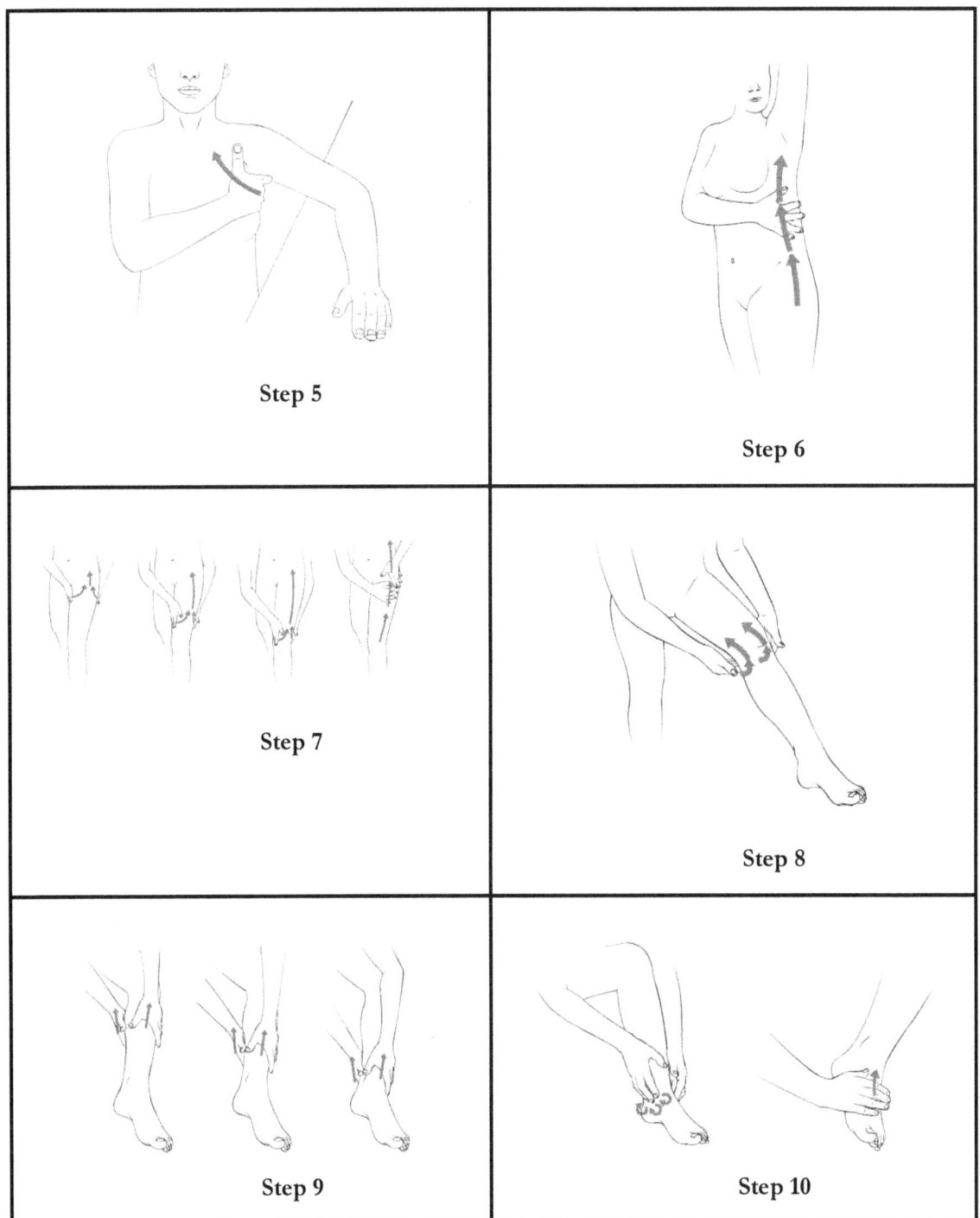

Step 5

Step 6

Step 7

Step 8

Step 9

Step 10

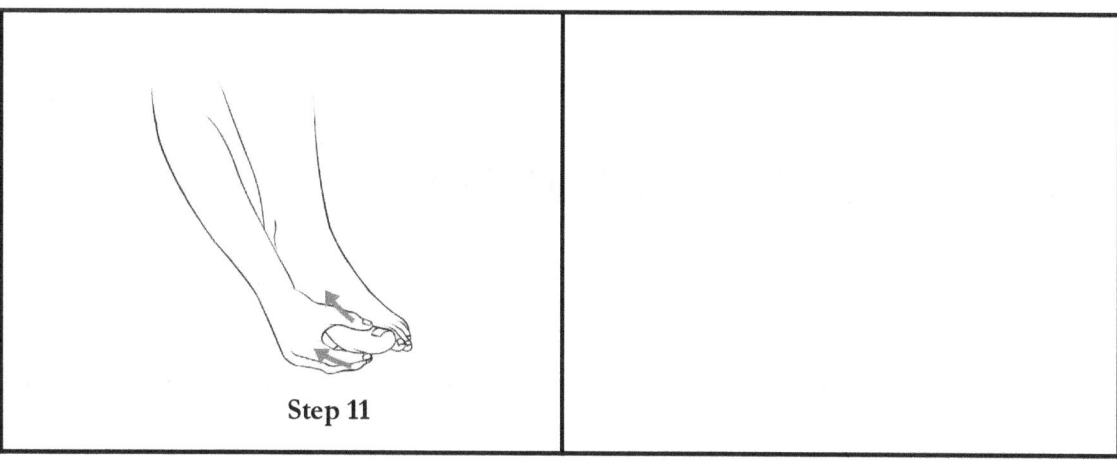

Step 11

EXERCISE 3.6
LYMPHATICS AND THE BREAST

Purpose of exercise: To examine lymphatics and the breasts.

Click on *Exercise 3.6* within your online platform or enter the address below into your web browser.
https://medlineplus.gov/ency/anatomyvideos/000084.htm

Please read the overview and watch the video on this website. Use the space below to write a brief summary of what you learned.

___
___
___
___
___
___
___
___
___
___
___
___
___
___

EXERCISE 3.7
LYMPH NODES

Purpose of exercise: To examine the role of lymphatic organs.

Click on *Exercise 3.7* within your online platform or enter the address below into your web browser.
https://medlineplus.gov/ency/anatomyvideos/000083.htm

Please read the overview and watch the video on this website. Use the space below to write a brief summary of what you learned.

_____
_____
_____
_____
_____
_____
_____
_____
_____
_____
_____
_____
_____
_____
_____

EXERCISE 3.8
SURGERIES OF LYMPHATIC STRUCTURES

Purpose of exercise: To explore various types of surgeries associated with lymphatic structures.

***Please make sure that your Adobe Flash Player is updated on your computer. Also, please be care not to click on any of the third-party advertisements because they will route you to another site.
You will perform the following virtual surgeries: **Adenoidectomy** and **Tonsillectomy.**

Enter the following addresses for each surgery type into your web browser. <u>Once you are ready to begin the surgery, click **START**. Please listen and read the instructions provided on the website and follow the interactive steps to perform the various surgeries.</u>

Click on *Exercise 3.8 (a)* within your online platform or enter the address below into your web browser.

Adenoidectomy: http://www.surgerysquad.com/surgeries/adenoidectomy/

Click on *Exercise 3.8 (b)* within your online platform or enter the address below into your web browser.
Tonsillectomy: http://www.surgerysquad.com/surgeries/tonsillectomy-surgery/

EXERCISE 3.9
SPLEEN

Purpose of exercise: To examine the role of lymphatic organs.

Click on *Exercise 3.9* within your online platform or enter the address below into your web browser.
http://www.wesnorman.com/spleen.htm

Please read the information provided on the website.

EXERCISE 3.10
IMMUNE RESPONSE

Purpose of exercise: To demonstrate the basic properties of immunity.

Click on *Exercise 3.10* within your online platform or enter the address below into your web browser.
https://medlineplus.gov/ency/anatomyvideos/000073.htm

Please read the overview and watch the video on this website. Use the space below to write a brief summary of what you learned.

_____
_____
_____
_____
_____
_____
_____
_____
_____
_____
_____
_____
_____
_____

EXERCISE 3.11
IMMUNITY

Purpose of exercise: To demonstrate the basic properties of immunity.

Click on *Exercise 3.11* within your online platform or enter the address below into your web browser.
https://www.nobelprize.org/educational/medicine/immunity/game/index.html

Please read and follow the instructions provided on the website.

EXERCISE 3.12
IMMUNE RESPONSES

Purpose of exercise: To demonstrate the basic properties of immunity.

Click on *Exercise 3.12* within your online platform or enter the address below into your web browser.
https://www.nobelprize.org/educational/medicine/immuneresponses/game/index.html#/plot1

Please read and follow the instructions provided on the website.

EXERCISE 3.13
ALLERGIES

Purpose of exercise: To describe significant immune disorders.

Click on *Exercise 3.13* within your online platform or enter the address below into your web browser.
https://medlineplus.gov/ency/anatomyvideos/000002.htm

Please read the overview and watch the video on this website. Use the space below to write a brief summary of what you learned.

_____
_____
_____
_____
_____
_____
_____
_____

_____
_____
_____
_____
_____

EXERCISE 3.14
ORDERING LABS - TRYPTASE

Purpose of exercise: To provide students with knowledge about common laboratory tests ordered by healthcare professionals to help diagnose cardiovascular medical conditions.

Click on *Exercise 3.14* within your online platform or enter the address below into your web browser.
https://labtestsonline.org/understanding/analytes/tryptase/tab/test/

Answer the questions about the lab test below using the hyperlink/website listed in this exercise. Summarize your answer so that is fits within the space provided.

1. Formal name:
2. Also known as:
3. How is it used?

4. When is it ordered?

5. What does the test result mean?

6. How is the sample collected for testing?

EXERCISE 3.15
ANTIBIOTIC INTERACTIVE QUIZ

Purpose of exercise: To identify medical methods used to diagnosis or treat immune disorders.

Click on *Exercise 3.15* within your online platform or enter the address below into your web browser.
https://www.cdc.gov/getsmart/community/about/quiz.html

Please read and follow the instructions provided on the website.

EXERCISE 3.16
THERAPEUTIC USE OF ANTIBODIES

Purpose of exercise: To identify medical methods used to diagnosis or treat immune disorders.

Click on *Exercise 3.16* within your online platform or enter the address below into your web browser.
https://youtu.be/hIEyVRghFX8

Please watch the animation on the website.

EXERCISE 3.17
CELLS OF THE IMMUNE SYSTEM

Purpose of exercise: To provide an overview of the immune system, concentrating on the roles played by B and T lymphocytes, and the antigen-presentation system.

Click on *Exercise 3.17* within your online platform or enter the address below into your web browser.
http://www.hhmi.org/biointeractive/cells-immune-system

Please read and watch the animation on the website.

EXERCISE 3.18
VACCINES

Purpose of exercise: To identify medical methods used to diagnosis or treat immune disorders.

Click on *Exercise 3.18* within your online platform or enter the address below into your web browser.
https://medlineplus.gov/ency/anatomyvideos/000137.htm

Please read the overview and watch the video on this website. Use the space below to write a brief summary of what you learned.

_____
_____
_____
_____
_____
_____
_____

EXERCISE 3.19
VACCINES

Purpose of exercise: To identify medical methods used to diagnosis or treat immune disorders.

Click on *Exercise 3.19* within your online platform or enter the address below into your web browser.
http://www.nhs.uk/Conditions/vaccinations/Pages/How-vaccines-work.aspx

Please read and watch the four animation on the website.

EXERCISE 3.20
ADULT VACCINE QUIZ – WHICH VACCINE DO YOU NEED?

Purpose of exercise: To identify medical methods used to diagnosis or treat immune disorders.

Click on *Exercise 3.20* within your online platform or enter the address below into your web browser.
https://www2a.cdc.gov/nip/adultimmsched/

Please read and watch the four animation on the website.

EXERCISE 3.21
PROTECTION OF THE BODY

Purpose of exercise: To provide an understanding about communicable diseases.

Click on *Exercise 3.21* within your online platform or enter the address below into your web browser.
http://lab.rockefeller.edu/steinman/interactive/dcell.html

Please read and watch the animation on the website.

EXERCISE 3.22

DETECTING BODY FLUIDS WITH A CELLPHONE BLACKLIGHT

Purpose of exercise: To demonstrate one method of detecting contaminated areas.

Black lights or UV lights are used by crime scene investigators to identify body fluids - including semen, sweat, saliva and urine. Semen happens to glow the brightest because of its particular mix of chemicals. This is because bodily fluids fluoresce - that means they absorb ultraviolet light and re-emit it as visible light. In this exercise you will trick your phone into becoming a blacklight to identify human body fluids in areas that appear "clean" to the naked eye. This should work for just about any smartphone, provide it has a flash.

Materials
- Smartphone with flash
- Smartphone flashlight app *(Free app)*
- Scotch Tape
- One blue Sharpie pen
- One purple Sharpie pen

Procedure
1. Cover the flash of your phone with a small piece of sticky tape.
2. Color over the tape with the blue Sharpie pen.
3. Add a second piece of tape over the top and color it with the purple pen.
4. Now turn on your flashlight app.
5. You now have a body fluid detecting device.

EXERCISE 3.23
BACTERIAL IDENTIFICATION VIRTUAL LAB

Purpose of exercise: To familiarize students with the science and techniques used to identify different types of bacteria based on their DNA sequences.

Click on *Exercise 3.23* within your online platform or enter the address below into your web browser.
http://media.hhmi.org/biointeractive/vlabs/bacterial_id/index.html?_ga=2.211562325.1125362674.1506365954-818520368.1502809801

Please read and follow the instructions provided on the website.

EXERCISE 3.24
WHAT KILLS GERMS VIRTUAL LAB

Purpose of exercise: To familiarize students with microbiology concepts such as inoculate petri dishes and observe zone of inhibition around substances such as bleach and antibiotics.

Click on *Exercise 3.24* within your online platform or enter the address below into your web browser.
http://www.glencoe.com/sites/common_assets/science/virtual_labs/LS08/LS08.html

Please read and follow the instructions provided on the website.

EXERCISE 3.25
IMMUNOLOGY VIRTUAL LAB

Purpose of exercise: Teaches the procedures of performing an ELISA test to determine whether a particular antibody is present in a patient's blood sample.

Click on *Exercise 3.25* within your online platform or enter the address below into your web browser.
http://media.hhmi.org/biointeractive/vlabs/immunology/index.html?_ga=2.52055913.1125362674.1506365954-818520368.1502809801

Please read and follow the instructions provided on the website.

EXERCISE 3.26
VIRTUAL PATHOLOGY

Purpose of exercise: To study and evaluate blood smears obtained from patients.

Click on *Exercise 3.26* within your online platform or enter the address below into your web browser.
http://glencoe.mheducation.com/sites/dl/free/0078802849/383963/BL_13.html

Please read and follow the instructions provided on the website.

EXERCISE 3.27
ORDERING LABS – ENA PANEL

Purpose of exercise: To provide students with knowledge about common laboratory tests ordered by healthcare professionals to help diagnose immunological medical conditions.

Click on *Exercise 3.27* within your online platform or enter the address below into your web browser.
https://labtestsonline.org/understanding/analytes/ena-panel/tab/test/

Answer the questions about the lab test below using the hyperlink/website listed in this exercise. Summarize your answer so that is fits within the space provided.

1. Formal name:
2. Also known as:
3. How is it used?

4. When is it ordered?

5. What does the test result mean?

6. How is the sample collected for testing?

EXERCISE 3.28
THE VIRTUAL AUTOPSY

Purpose of exercise: To teach students how to think critically about information they have learned and how to use it to answer questions about a person's illness or cause of death in case scenarios

Click on *Exercise 3.28* within your online platform or enter the address below into your web browser.
http://www.le.ac.uk/pa/teach/va/titlpag1.html

Please complete the following **Cases: 3, 11, and 16**. Read and follow the instructions provided on the website.

EXERCISE 3.29
REVIEW QUESTIONS

Please continue by answering the questions below.

LYMPHATIC
1. Water within capillaries is called _____.
2. Tissue fluid is made from _____ by the process of _____.
3. Tissue fluid that has entered lymph capillaries is called _____.
4. Backflow of lymph in the larger lymph vessels is prevented by _____.
5. Lymph is kept moving in the larger lymph vessels by contraction of the _____ in their walls.
6. In the larger lymph vessels of the legs, lymph is kept moving by the _____.
7. The _____ lymph nodes destroy pathogens in the lymph returning from the arms.
8. The _____ lymph nodes destroy pathogens in the lymph returning from the head.
9. The _____ lymph nodes destroy pathogens in the lymph returning from the head.

10. As a person reaches adulthood, the thymus gland _____ in size.

IMMUNITY

11. Antigens that are found on the cells of an individual are called _____ antigens.
12. Foreign antigens are those that will stimulate production of _____.
13. The component of immunity that is specific as to antigen is _____ immunity.
14. The component of immunity that is not specific as to antigen is _____ immunity.
15. The component of immunity that creates memory is _____ immunity.
16. The component of immunity that does not create memory is _____ immunity.
17. The component of immunity that may become more efficient with repeated exposures is _____ immunity.
18. In innate immunity, the body's outermost defense is an unbroken _____.
19. The respiratory mucosa is lined with _____ to sweep inhaled pathogens out.
20. The cells of innate immunity that produce histamine and leukotrienes are the _____ and _____.
21. In innate immunity, the function of interferon is to prevent the reproduction of _____ within cells.
22. In innate immunity, the chemical that lyses cellular antigens or labels non-cellular antigens is _____.
23. In innate immunity, the signs of inflammation are pain, redness, _____, and _____.
24. Inflammation is the body's response to _____, and is part of _____ immunity.
25. In adaptive immunity, macrophages and helper T cells work together to _____.
26. In adaptive immunity, the _____ T cells chemically destroy foreign antigens.
27. In adaptive immunity, when antibodies bond to viruses, they change the shape of the viruses, which is called _____.
28. A vaccine may contain a(n) _____ or a(n) _____ as an antigen.
29. A vaccine stimulates production of _____ and _____.
30. A vaccine works because it takes the place of the _____ to the pathogen.
31. Recovery from a disease provides _____ acquired _____ immunity.
32. A vaccine such as that for measles provides _____ acquired _____ immunity.
33. An injection of gamma globulins provides _____ acquired _____ immunity.
34. Artificially acquired active immunity occurs when a person _____.
35. Naturally acquired active immunity occurs when a person _____.

THE RESPIRATORY SYSTEM
LAB 4

CRASHCOURSE VIDEO(S):

Click on the video embedded within your online platform or enter the address below into your web browser:
1. **https://youtu.be/bHZsvBdUC2I**
2. **https://youtu.be/Cqt4LjHnMEA**

(Please make sure to watch the video before continuing)

DEFINING KEY TERMS:

1. Alveolar Ventilation:

2. Anatomic Dead Space (ADS):

3. Apnea:

4. Asthma:

5. Atmospheric Pressure:

6. Auscultating:

7. Bronchoscopy:

8. Crackles (Rales):

9. Expiration:

10. Expiratory Reserve Volume (ERV):

11. Functional Residual Capacity:

12. Hypoxia:

13. Inspiration:

14. Inspiratory Capacity:

15. Inspiratory Reserve Volume (IRV):

16. Minute Ventilation (MV):

17. Oxyhemoglobin:

18. Paranasal Sinuses:

19. Partial Pressure:

20. Pleural Membranes:

21. Residual Volume (RV):

22. Rhinoplasty:

23. Spirometer:

24. Stridor:

25. Surfactant:

26. Tidal Volume:

27. Total Lung Capacity:

28. Tuberculosis:

29. Type I Alveolar cells:

30. Type II Alveolar cells:

31. Ventilation:

32. Vital Capacity (VC):

33. Wheezes:

EXERCISE 4.1
IDENTIFYING RESPIRATORY TISSUE UNDER THE MICROSCOPE

Purpose of exercise: To observe examples of respiratory tissue types under the microscope

Click on *Exercise 4.1 (a)* within your online platform or enter the address below into your web browser:
http://histologyguide.org/

Once you enter the site, click onto the **Slide Box** tab located on the left hand side. You will see a list of slide categorized by tissue type and organ system in bold font. Click onto the tabs that correctly identifies the tissue type you must observe for this exercise. The tissue type is listed on the table below.

If you encounter difficulties linking to this web address, or would like to view a different source try the link below.

Click on *Exercise 4.1 (b)* within your online platform or enter the address below into your web browser:
http://www.kumc.edu/instruction/medicine/anatomy/histoweb/index.htm

(Your instructor may choose to upload pictures of the tissue types in your college's online platform. If this is the case, you can still utilize the listed slides below as a reference to identify.)

Make a drawing of each prepared slide on high power in the circles below.

Prepared Slide
1. Nasal Conchae and Palate
2. Epiglottis
3. Larynx
4. Trachea
5. Lung

Observations

Tissue Type	Tissue Drawing on High Power (40X Objective)
• Nasal Conchae and Palate	
• Epiglottis	

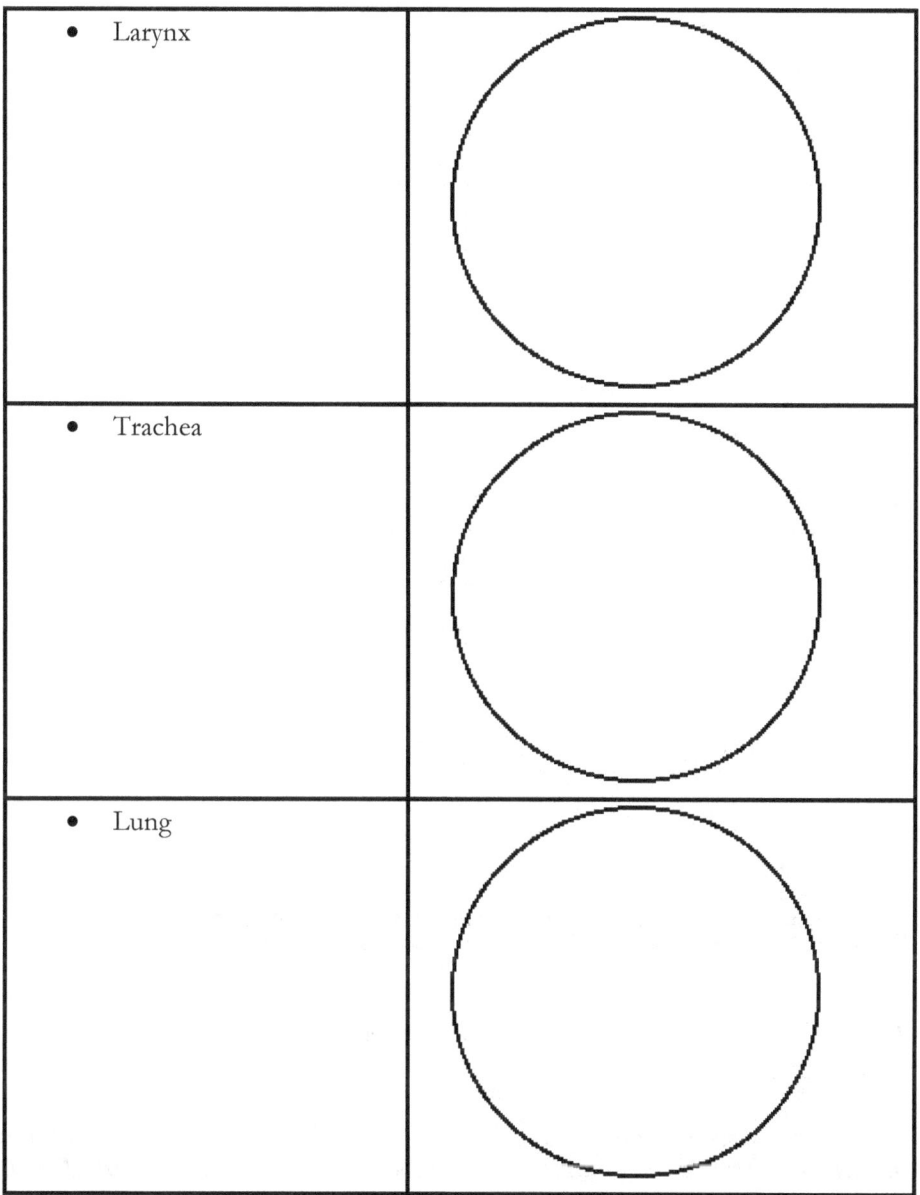

- Larynx

- Trachea

- Lung

EXERCISE 4.2
BREATHING

Purpose of exercise: To provide an overview of the breathing mechanism.

Click on *Exercise 4.2* within your online platform or enter the address below into your web browser.
https://medlineplus.gov/ency/anatomyvideos/000018.htm

Please read the overview and watch the video on this website. Use the space below to write a brief summary of what you learned.

___

_____
_____
_____
_____
_____
_____
_____
_____
_____
_____
_____
_____
_____
_____
_____

EXERCISE 4.3
ANATOMICAL PARTS

Purpose of exercise: To examine the structure associated with the organ system.

Click on *Exercise 4.3* within your online platform or enter the address below into your web browser:
https://www.biodigital.com/

- *(This is a free site, but you will need to sign up with your name and a validated email address. You can use your personal email address or school email address. MAKE SURE TO CREATE A PASSWORD YOU CAN REMEMBER. I WOULD SUGGEST WRITING IT DOWN AND KEEPING IT IN A SAFE PLACE. YOU WILL USE IT AGAIN.)*

Click **SIGN UP**. Once you have provided the appropriate information, you are now able to access BioDigital's website content. Click **LOG IN** using the email and password you created. Then click **SIGN IN**. On the left of your screen choose from the systems listed.

Please click on the square boxes next to this icon. Now click on the **Anatomy By Systems** icon. Click on the following image(s) with caption. **Male Respiratory System, Female Respiratory System.** View the structures associated with this system.

EXERCISE 4.4

PHARYNX DISSECTION

Purpose of exercise: To describe the anatomy of the pharynx through dissection.

Click on *Exercise 4.4* within your online platform or enter the address below into your web browser.
http://act.downstate.edu/courseware/haonline/labs/L31/ld0000.htm

Please read the information provided on the website. Click on each step to view the dissection.

EXERCISE 4.5
FINDING YOUR WAY AROUND THE THORAX

Purpose of exercise: To describe the anatomy of the thorax.

Click on *Exercise 4.5* within your online platform or enter the address below into your web browser.
http://www.wesnorman.com/thoraxlesson2.htm

Please read the information provided on the website.

EXERCISE 4.6
PLEURAL CAVITIES AND LUNGS DISSECTION

Purpose of exercise: To describe the anatomy of the thorax through dissection.

Click on *Exercise 4.6* within your online platform or enter the address below into your web browser.
http://act.downstate.edu/courseware/haonline/labs/L19/ld0000.htm

Please read the information provided on the website. Click on each step to view the dissection.

EXERCISE 4.7
RADIOLOGICAL INTERPRETATION

Purpose of exercise: To provide students a foundational knowledge in the art of radiological interpretation while reinforcing anatomy and physiology concepts.

During a radiographic procedure, an x-ray beam is passed through the body. A portion of the x-rays are absorbed or scattered by the internal structure and the remaining x-ray pattern is transmitted to a detector so that an image may be recorded for later evaluation. The recoding of the pattern may occur on film or through electronic means. X-rays are a type of radiation called electromagnetic waves. X-ray images show the parts of your body in different shades of black (radiolucent) and white (radiopaque). This is because different tissues absorb different amounts of radiation. Calcium in bones absorbs x-rays the most, so bones look white. Fat and other soft tissues absorb less, and look gray. Air absorbs the least, so lungs look black.

The most familiar use of x-rays is checking for broken bones, but x-rays are also used in other ways. For example, chest x-rays can spot pneumonia. Mammograms use x-rays to look for breast cancer. It is used to diagnose or treat patients by recording images of the internal structure of the body to assess the presence or absence of disease, foreign objects, and structural damage or anomaly.

This website allows students to build their skills in medical x-ray imaging by giving free access to materials. **YOU ARE NOT REQUIRED TO BUY ANYTHING ON THE RADIOLOGYMASTER CLASS WEBSITE FOR THIS COURSE**.

Click on *Exercise 4.7(a)* within your online platform or enter the address below into your web browser before starting this exercise: https://www.radiologymasterclass.co.uk/tutorials/tutorials_howto

Now click on *Exercise 4.7(b)* within your online platform or enter the address below into your web browser: https://www.radiologymasterclass.co.uk/tutorials/chest/chest_home_anatomy/chest_anatomy_start

Chest X-ray Anatomy
This tutorial demonstrates some of the important anatomical structures visible on a chest X-ray. These structures are discussed in a specific order to help you develop your own systematic approach to viewing chest X-rays. By the end of the tutorial you will be familiar with all the important structures of the chest, which should be checked whenever you look at a chest X-ray.

A systematic approach for viewing chest X-rays ensures no important structures are ignored, but a flexible approach is required to suit each clinical setting. This tutorial demonstrates some of the important anatomical structures visible on a chest X-ray. These structures are discussed in a specific order (**trachea and bronchi, Hilar structures, lung zones, pleura, lung lobes and fissures, costophrenic angles, diaphragm, heart, mediastinum, soft tissues, and bones**) to help you develop your own systematic approach to viewing chest X-rays. By the end of the tutorial you will be familiar with all the important structures of the chest, which should be checked whenever you look at a chest X-ray. Before you start, look at the normal chest X-ray. Please read and follow the instructions provided on the website. Click the "**Next >>**" button to move through the images on the page.

EXERCISE 4.8
VENTILATION

Purpose of exercise: To provide an overview of the breathing mechanism.

Click on *Exercise 4.8* within your online platform or enter the address below into your web browser. http://www.wiley.com/college/powerphysdemo/section8/animation01/index.html

Please read and watch the animation on the website.

EXERCISE 4.9
RESPIRATORY MOVEMENT

Purpose of exercise: To provide an overview of the breathing mechanism.

Click on *Exercise 4.9* within your online platform or enter the address below into your web browser.

http://www.wesnorman.com/respiratorymovements.htm

Please read the information provided on the website.

EXERCISE 4.10
IDENTIFYING PARTS OF THE RESPIRATORY SYSTEM VIRTUAL LAB

Purpose of exercise: To allows students to observe and interact with a graphic of the respiratory system.

Click on *Exercise 4.10* within your online platform or enter the address below into your web browser.
http://www.glencoe.com/sites/common_assets/science/virtual_labs/LS24/LS24.html

Please read and follow the instructions provided on the website.

EXERCISE 4.11
INTERPRETATION OF ABGS (ACID-BASE)

Purpose of exercise: To demonstrate the role of acid-base regulation.

Click on *Exercise 4.11* within your online platform or enter the address below into your web browser.
http://fitsweb.uchc.edu/student/selectives/TimurGraham/Acid_Base_Physiology.html

Please read the information provided on the website. Click on the four teaching modules within this site.

EXERCISE 4.12
VIRTUAL BRONCHOSCOPY SIMULATION

Purpose of exercise: To demonstrate to student how healthcare professionals move the bronchoscope throughout the tracheo-bronchial tree.

Click on *Exercise 4.12* within your online platform or enter the address below into your web browser.
http://pie.med.utoronto.ca/VB/VB_content/assets/applications/index.html

Please read and follow the instructions provided on the website. Students are able to move the bronchoscope throughout the tracheo-bronchial tree by clicking on the directional arrows under the "Bronchoscopic view" at the right of the simulation. Clicking on the labels (A-Z) on the "Bronchoscopic view" provides details of the structures. Navigation is aided by the "Bronchial Tree Navigational Map" on the left of the simulation, which shows the current location of the bronchoscope as an orange line in the airway.

EXERCISE 4.13
INTRODUCTION TO SPIROMETERS

Purpose of exercise: To demonstrate how Pulmonary Function Tests (PFT) are used to check the health of the lungs and respiratory passageways.

Click on *Exercise 4.13* within your online platform or enter the address below into your web browser.
https://www.getbodysmart.com/ap/respiratorysystem/physiology/spirometry/spirometer/tutorial.html

Please read and follow the instructions provided on the website. Make sure to click the **blue links** for labels and animations.

EXERCISE 4.14
PULMONARY FUNCTION TESTS USING SPIROMETRY

Purpose of exercise: To demonstrate how Pulmonary Function Tests (PFT) are used to check the health of the lungs and respiratory passageways.

Click on *Exercise 4.14* within your online platform or enter the address below into your web browser.
https://www.getbodysmart.com/ap/respiratorysystem/physiology/spirometry/pulmonary_function/tutorial.html

Please read and follow the instructions provided on the website. Make sure to click the **blue links** for labels and animations.

EXERCISE 4.15
ASTHMA

Purpose of exercise: To describe significant respiratory disorders.

Click on *Exercise 4.15* within your online platform or enter the address below into your web browser.
http://learn.genetics.utah.edu/content/history/asthma/

Please read and watch the animation on the website.

EXERCISE 4.16
ASTHMA RISK ASSESSMENT

Purpose of exercise: To describe significant respiratory disorders

Click on *Exercise 4.16* within your online platform or enter the address below into your web browser.
http://www.aaaai.org/conditions-and-treatments/asthma/asthma-quiz

Please read and follow the instructions provided on the website

EXERCISE 4.17
TUBERCULOSIS

Purpose of exercise: To describe significant respiratory disorders.

Click on *Exercise 4.17* within your online platform or enter the address below into your web browser.
https://www.nobelprize.org/educational/medicine/tuberculosis/tbc/index.html

Please read and follow the instructions provided on the website.

EXERCISE 4.18
RESPIRATORY OR LUNG VOLUMES AND CAPACITIES

Purpose of exercise: To exhibit the various factors that control the rate of respiration.

Click on *Exercise 4.18* within your online platform or enter the address below into your web browser.
https://www.getbodysmart.com/ap/respiratorysystem/physiology/spirometry/volumes_capacities/animation.html

Please read and follow the instructions provided on the website. Make sure to click the **blue links** for labels and animations.

EXERCISE 4.19
SLEEP APNEA SCREENING QUESTIONNAIRE

Purpose of exercise: To describe significant respiratory disorders.

Sleep apnea is a condition in which a person stops breathing periodically during sleep. These cessations in breathing can occur anywhere from a few times a night up to hundreds of times a night. The result of this interrupted breathing pattern is severely fragmented sleep, as the individual must wake up enough to regain muscle control in the throat and to reopen the airway. This constant awakening means that people with apnea do not get sufficient or good quality sleep, resulting in sleepiness and/or fatigue. But, because **Obstructive Sleep Apnea (OSA)** sufferers typically do not gain full consciousness when they wake after apnea episodes, they often do not know the cause of their sleepiness and/or fatigue. Along with sleepiness and/or fatigue, **OSA** a can cause significant physiological and psychological distress.

The National Healthy Sleep Awareness Project warns that untreated, severe obstructive sleep apnea hurts HEARTS by increasing the risk of:

H – Heart failure
E – Elevated blood pressure

A – Atrial fibrillation
R – Resistant hypertension
T – Type 2 diabetes
S – Stroke

Fortunately, treatments for OSA are available. Following diagnosis by a board-certified sleep medicine physician, the most commonly prescribed treatment for sleep apnea is Continuous positive airway pressure therapy (CPAP).

DO I HAVE SLEEP APNEA?
SCREENING QUESTIONNAIRE

This sleep apnea screener features the **STOP BANG** questionnaire and **Epworth Sleepiness Scale** to help you gauge your risk for sleep apnea. Get a pencil and paper ready to write down your answers to each question.

STOP BANG	EPWORTH (RATE 0 – 3 FOR EACH SCENARIO)

(Answer yes or no for each question)
 	
- **S (snore)**
  1. Do you snore?
- **T (tired)**
  1. Do you feel fatigued during the day?
  2. Do you wake up feeling like you haven't slept?
- **O (obstruction)**
  1. Have you been told you stop breathing at night?
  2. Do you gasp for air or choke while sleeping?
- **P (pressure)**
  1. Do you have high blood pressure or are on medication to control high blood pressure?
 	
SCORE: If you checked **YES** to two or more questions on the **STOP** portion you are at risk for OSA.
 	
- **B (BMI)**
  1. Is your body mass index greater than 28?
- **A (age)**
  1. Are you 50 years old or older?
- **N (neck)**
  1. Are you a male with neck circumference greater than 17 inches, or a female with neck circumference greater than 16 inches?
- **G (gender)**
  1. Are you a male?
 	
SCORE: The more questions you checked **YES** to on the **BANG** portion, **the greater your risk** of having moderate to severe Obstructive Sleep Apnea.
	How likely are you to doze off or fall asleep in the situations described below, in contrast to feeling just tired? This refers to your usual way of life in recent times. Even if you haven't done some of these things recently, try to work out how they would have affected you. Use the following scale to choose the most appropriate number for each situation:
	
	0 = Would never doze
	1 = Slight chance of dozing
	2 = Moderate chance of dozing
	3 = High chance of dozing
	- Sitting and reading
	- Watching TV
	- Sitting inactive in a public place (e.g. a theater or a meeting)
	- Sitting in a car as a passenger for a continuous hour
	- Lying down to rest in the afternoon when circumstances permit
	- Sitting and talking to someone
	- Sitting quietly after a lunch without alcohol
	- Sitting in a car stopped in traffic for a few minutes
	
	SCORE: Add up your score for each scenario. **0–10** Normal range | **10–12** Borderline | **12–24** Sleepy

EXERCISE 4.20
BREATH SOUNDS REFERENCE GUIDE WITH AUDIO
(AUSCULTATING FOR BREATH SOUNDS)

Purpose of exercise: To exhibit normal and abnormal respiratory sounds.

Click on *Exercise 4.20* within your online platform or enter the address below into your web browser.
https://www.practicalclinicalskills.com/breath-sounds-reference-guide

Please read and follow the instructions provided on the website. Click on any of the titles to review text, waveforms and audio recordings.

EXERCISE 4.21
EFFECTS OF USE OF TOBACCO

Purpose of exercise: To demonstrate the negative health effects of tobacco usage.

In this exercise you will experience the lack of oxygen in the lungs caused by smoking and determine the amount of tar the lungs accumulate from tobacco use. You will see how the tissues of the lungs become hard when you are a smoker. The poisons in tobacco are absorbed through the skin in the mouth and through the lungs causing body functions to slow down or stop. One of the poisons in tobacco is tar. It coats the air sacs in the lungs until you can no longer breath and you suffocate as seen in the condition emphysema.. Over the years you are slowly suffocating, your body does not get the amount of oxygen it needs and the blood vessels get narrow. Soon a simple activity like walking is hard to do.

Materials:
- Drinking Straws
- Two Sponges
- Water

Procedure One:
1. Cut a drinking straw three-inches.
2. Run in place for 1 minute
3. Put the straw in your mouth and breathe only through the straw (not through the nose)
4. Bite gently on the straw as you try to breathe to simulate an even more severe case of emphysema (slowly suffocating to death)
5. Resume normal breathing without the straw.

Procedure Two:
1. Allow participants to compare the feel of a damp, soft sponge and a dry, hard sponge.
2. The sponges demonstrate the difference between healthy lung tissue and damaged tissue.

Question
Why is hardened lung tissue a problem?

EXERCISE 4.22
BREATHING DIFFICULTY CHECKUP

Purpose of exercise: To describe significant respiratory disorders.

Click on *Exercise 4.22* within your online platform or enter the address below into your web browser.
http://www.freemd.com/breathing-difficulty/overview.htm

Please read and follow the instructions provided on the website.

EXERCISE 4.23
RHINOPLASTY SURGERY

Purpose of exercise: To explore cosmetic rhinoplasty surgery.

***Please make sure that your Adobe Flash Player is updated on your computer. Also, please be care not to click on any of the third-party advertisements because they will route you to another site.
You will perform the following virtual surgery: **Rhinoplasty**

Enter the following addresses for each surgery type into your web browser. <u>Once you are ready to begin the surgery, click **START**. Please listen and read the instructions provided on the website and follow the interactive steps to perform the various surgeries.</u>

Click on *Exercise 4.23 (a)* within your online platform or enter the address below into your web browser.
Rhinoplasty: http://www.surgerysquad.com/surgeries/virtual-nose-job-virtual-rhinoplasty/

EXERCISE 4.24
ORDERING LABS – AFB TESTING

Purpose of exercise: To provide students with knowledge about common laboratory tests ordered by healthcare professionals to help diagnose respiratory medical conditions.

Click on *Exercise 4.24* within your online platform or enter the address below into your web browser.
https://labtestsonline.org/understanding/analytes/afb-culture/tab/test/

Answer the questions about the lab test below using the hyperlink/website listed in this exercise. Summarize your answer so that is fits within the space provided.

1. Formal name:
2. Also known as:
3. How is it used?

4. When is it ordered?

5. What does the test result mean?

6. How is the sample collected for testing?

EXERCISE 4.25
THE VIRTUAL AUTOPSY

Purpose of exercise: To teach students how to think critically about information they have learned and how to use it to answer questions about a person's illness or cause of death in case scenarios.

Click on *Exercise 4.25* within your online platform or enter the address below into your web browser.
http://www.le.ac.uk/pa/teach/va/titlpag1.html

Please complete the following **Cases: 5, 7, 10, and 12**. Read and follow the instructions provided on the website.

EXERCISE 4.26
REVIEW QUESTIONS

Please continue by answering the questions below.

RESPIRATORY

1. The parts of the pharynx that are passageways for both air and food are the _____ and the _____.
2. During swallowing, the larynx is covered by the _____.
3. The trachea is lined with _____ epithelial tissue.
4. The trachea is kept open by _____ in the shape of a(n) _____.
5. The trachea and all of the respiratory passages are collectively called the _____.
6. The bronchioles differ from the bronchi in that there is no _____ in their walls and they may _____.
7. The right primary bronchus branches into _____ secondary bronchi, and the left primary bronchus branches into _____.
8. The _____ pleura covers the lungs and the _____ pleura lines the chest cavity.
9. The alveoli and the pulmonary capillaries are all made of _____ tissue, which permits _____.
10. Pulmonary surfactant permits normal inhalation because it decreases the _____ within alveoli.

11. The diaphragm is made of _____ muscle, and during inhalation it moves _____.
12. The diaphragm _____ during inhalation and moves _____.
13. The diaphragm _____ during exhalation, and moves _____.
14. The diaphragm contracts during _____ and relaxes during _____.
15. The external intercostal muscles contribute to _____ by pulling the ribs _____.
16. The internal intercostal muscles contribute to a(n) _____ by pulling the ribs _____.
17. In external respiration, oxygen diffuses from the _____ to the _____.
18. In external respiration, carbon dioxide diffuses from the _____ to the _____.
19. In external respiration, the PO_2 in the alveoli is _____, and the PO_2 in the surrounding blood is _____.
20. In external respiration, the PCO_2 in the alveoli is _____, and the PCO_2 in the surrounding blood is _____.
21. The partial pressure of a gas in air or a body fluid is a measure of the _____ of the gas.
22. In internal respiration, the PO_2 in the blood is _____, and the PO_2 in the tissues is _____.
23. In internal respiration, the PCO_2 in the blood is _____, and the PCO_2 in the tissues is _____.
24. Most oxygen is transported to tissues bonded to the _____ in _____.
25. The mineral that is essential for oxygen transport is _____, because it is part of _____.
26. Oxygen is released from hemoglobin when the PCO_2 of the surrounding tissues is _____.
27. The CNS respiratory centers are located in the _____ and _____.
28. The CNS respiratory centers are located in the _____ and _____.
29. Severe respiratory acidosis will cause the blood pH to fall below _____.

THE DIGESTIVE SYSTEM
LAB 5

CRASHCOURSE VIDEO(S):

Click on the video embedded within your online platform or enter the address below into your web browser:
1. **https://youtu.be/yIoTRGfcMqM**
2. **https://youtu.be/pqgcEIaXGME**
3. **https://youtu.be/jGme7BRkpuQ**

(Please make sure to watch the video before continuing)

DEFINING KEY TERMS:

1. Alcohol:
2. Bile:
3. Cholecystokinin:
4. Colon:
5. Constipation:
6. Defecation:
7. Digestion:
8. Enamel:
9. Epiglottis:
10. Feces:
11. Gallbladder:
12. Gastric Bypass:
13. Heart burn:
14. Liver:
15. Pancreas:

16. Peristalsis:

17. Pharynx:

18. Stomach:

19. Ulcer:

EXERCISE 5.1
IDENTIFYING GASTROINTESTINAL TISSUES UNDER THE MICROSCOPE

Purpose of exercise: To identify the general histological layers of the digestive organs and explain how the layers are modified to accommodate the function of each organ.

Click on *Exercise 5.1 (a)* within your online platform or enter the address below into your web browser:
http://histologyguide.org/

Once you enter the site, click onto the **Slide Box** tab located on the left hand side. You will see a list of slide categorized by tissue type and organ system in bold font. Click onto the tabs that correctly identifies the tissue type you must observe for this exercise. The tissue type is listed on the table below.

If you encounter difficulties linking to this web address, or would like to view a different source try the link below.

Click on *Exercise 5.1 (b)* within your online platform or enter the address below into your web browser:
http://www.kumc.edu/instruction/medicine/anatomy/histoweb/index.htm

(Your instructor may choose to upload pictures of the tissue types in your college's online platform. If this is the case, you can still utilize the listed slides below as a reference to identify.)

Make a drawing of each prepared slide on high power in the circles below.

Prepared Slide *(Tissue Example)*
1. Hard Palate
2. Tongue
3. Esophagus
4. Cardiac Stomach
5. Fundic Stomach
6. Pyloric Stomach
7. Duodenum
8. Jejunum
9. Ileum
10. Colon
11. Appendix
12. Rectum
13. Liver
14. Gallbladder

Observations

Tissue Type	Tissue Drawing on High Power (40X Objective)
1. Hard Palate	
2. Tongue	
3. Esophagus	
4. Cardiac Stomach	

5. Fundic Stomach

6. Pyloric Stomach

7. Duodenum

8. Jejunum

9. Ileum	
10. Colon	
11. Appendix	
12. Rectum	

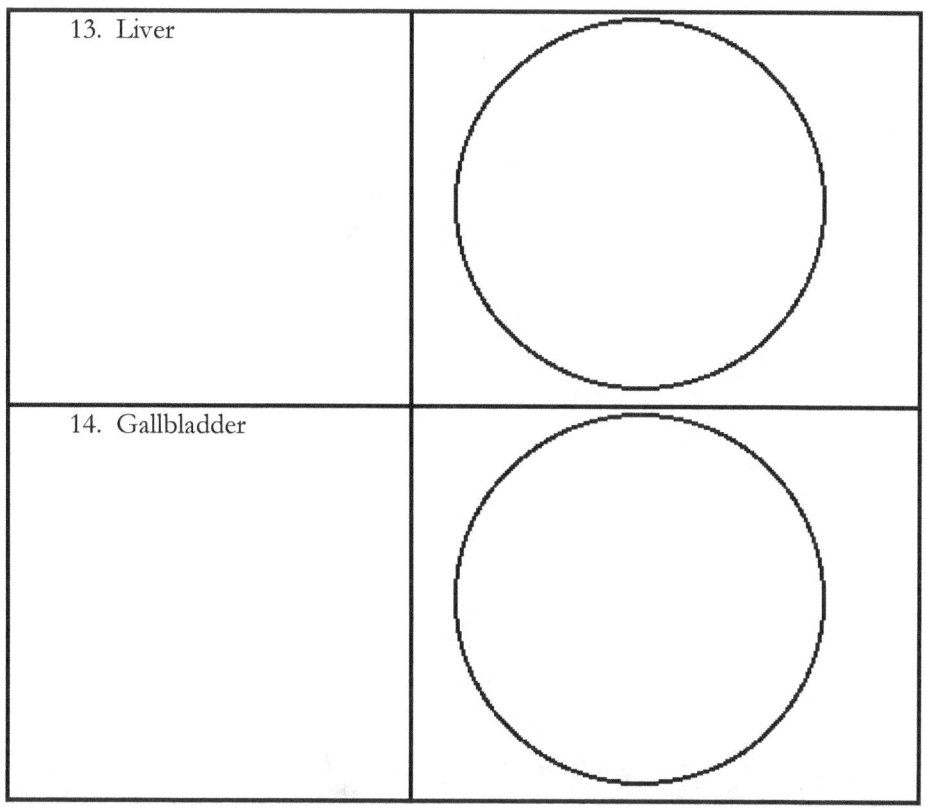

EXERCISE 5.2
ASSESSING TOOTH ENAMEL

Purpose of exercise: To demonstrate the effect soda has on tooth enamel and decay.
(NOTE: This experiment is expected to take one week before you get to see and assess the results.)

Enamel is the thin outer covering of the tooth. This tough shell is the hardest tissue in the human body. Tooth enamel is the hard outer substance of the tooth that protects the softer and more sensitive inner tooth, called dentin. If the enamel is exposed to destructive bacteria bred from sugars and starches, or to acids from citrus fruits and coffee, for example, the substance starts to break down. Soda may taste sweet and refreshing, but it's bad news for your teeth's enamel in large amounts. Most sodas are full of sugar, which contributes to the production of decay-causing bacteria. Even diet soda or unsweetened fizzy drinks, can lead to tooth enamel loss because they are so acidic. This is why it is so important to properly brush and floss your teeth at least twice each day to make sure that these harmful substances don't stay in contact with your tooth enamel for too long.

Materials
- 1 small bottle each of Coca Cola, Pepsi, Dr. Pepper, Sprite, Mountain Dew and Distilled water
- 6 plastic cups
- 6 tarnished pennies
- Notepad
- Measuring cup
- Marker for creating cup labels

Procedure

1. Take all the 6 plastic cups and label each using the marker. Assign one cup for each drink - one for Coca Cola, one for Pepsi, one for Dr. Pepper, one for Sprite, one for Mountain Dew and the last one for the distilled water.
2. Pour each liquid into the designated cup and drop one tarnished penny for each labelled cup.
3. Observe what happens to each penny each day. Take note and record these observations in your notepad. *(You may pick the penny out of the cup to look at them closely but make sure you return them back inside after observing.)*
4. When noting your observations, try to observe whether the darker colored sodas remove the tarnish from the pennies faster than the lighter-colored ones.
5. Also take note if the lighter colored sodas changed color as they remove the tarnish.
6. Do not forget to compare these observations to that of the penny soaked in the distilled water.
7. Continue doing this assessment for a week.

Question

Which soda was more corrosive and would be more damaging to the human tooth enamel?

EXERCISE 5.3
OVERVIEW OF THE DIGESTIVE SYSTEM

Purpose of exercise: To identify the organs of the gastrointestinal tract and the accessory organs of digestion and their functions in digestion.

Click on *Exercise 5.3* within your online platform or enter the address below into your web browser.
http://linkstudio.info/animation/GI-Digestive-System.html

Please read and watch the animation on the website.

EXERCISE 5.4
ANATOMICAL PARTS

Purpose of exercise: To identify the organs of the gastrointestinal tract and the accessory organs of digestion and their functions in digestion.

Click on *Exercise 5.4* within your online platform or enter the address below into your web browser:
https://www.biodigital.com/

- *(This is a free site, but you will need to sign up with your name and a validated email address. You can use your personal email address or school email address. MAKE SURE TO CREATE A PASSWORD YOU CAN REMEMBER. I WOULD SUGGEST WRITING IT DOWN AND KEEPING IT IN A SAFE PLACE. YOU WILL USE IT AGAIN.)*

Click **SIGN UP**. Once you have provided the appropriate information, you are now able to access BioDigital's website content. Click **LOG IN** using the email and password you created. Then click **SIGN IN**. On the left of your screen choose from the systems listed.

Please click on the square boxes next to this icon. ⊞ [Search Human Library] . Now click on the **Anatomy By Systems** icon. [▯] Click on the following image(s) with caption. **Male Digestive System, Female Digestive System.** View the structures associated with this system.

EXERCISE 5.5
DIGESTIVE SYSTEM CONCEPT MAP

Purpose of exercise: To identify the organs of the gastrointestinal tract and the accessory organs of digestion and their functions in digestion.

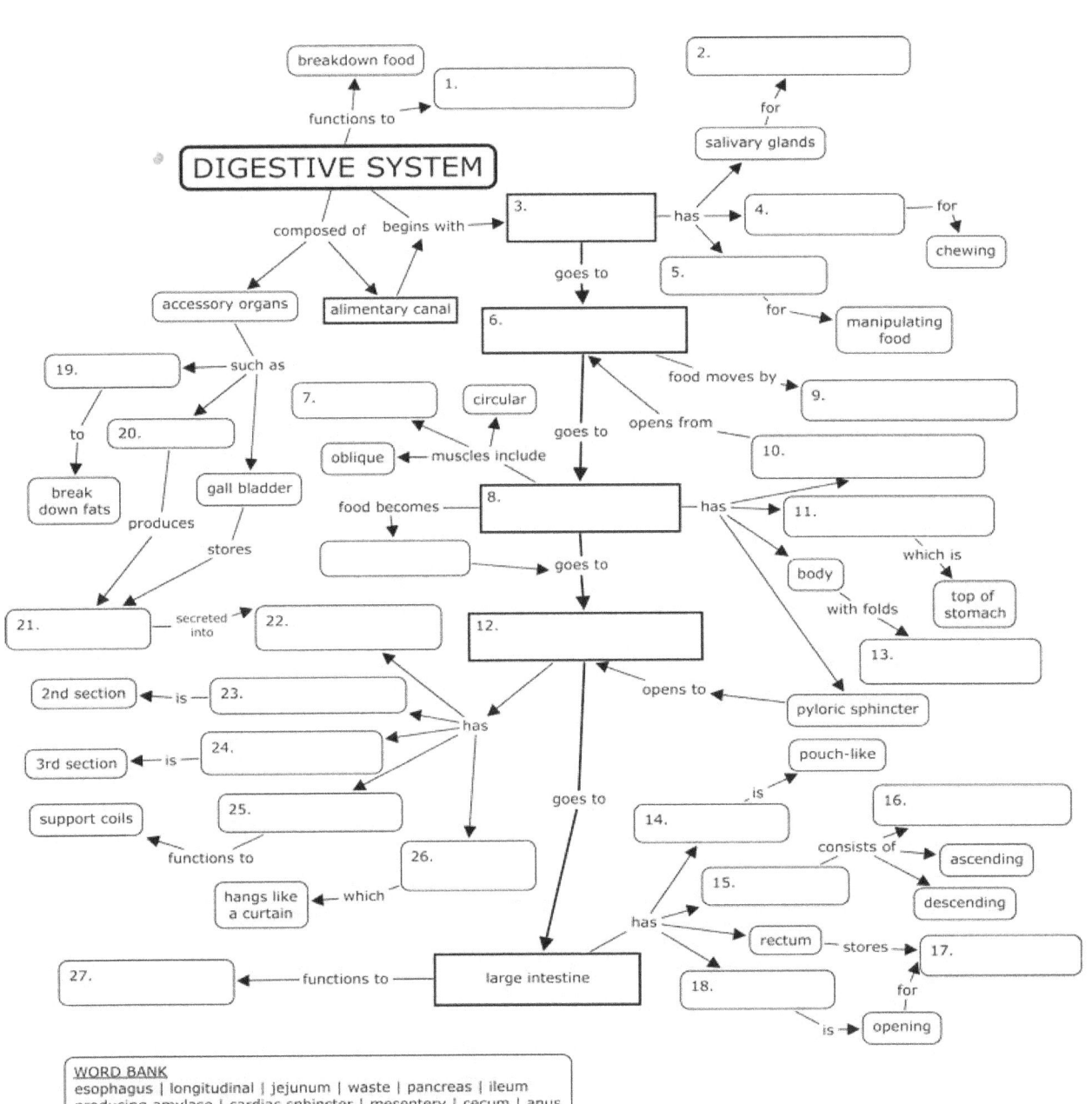

WORD BANK
esophagus | longitudinal | jejunum | waste | pancreas | ileum
producing amylase | cardiac sphincter | mesentery | cecum | anus
mouth | rugae | peristalsis | tongue | transverse | small intestine
teeth | absorb nutrients | stomach | greater omentum | bile
fundus | duodenum | chyme | absorb water | colon | liver

126

EXERCISE 5.6
THE ABDOMINAL CAVITY AND ITS CONTECTS

Purpose of exercise: To describe the anatomy of the abdominal cavity through dissection.

Click on *Exercise 5.6* within your online platform or enter the address below into your web browser.
http://www.wesnorman.com/abdominalcavity.htm

Please read the information provided on the website.

EXERCISE 5.7
ABDOMINAL CAVITY DISSECTION

Purpose of exercise: To describe the anatomy of the abdominal cavity through dissection.

Click on *Exercise 5.7* within your online platform or enter the address below into your web browser.
http://act.downstate.edu/courseware/haonline/labs/L37/ld0000.htm

Please read the information provided on the website. Click on each step to view the dissection.

EXERCISE 5.8
DIGESTION

Purpose of exercise: To demonstrate the mechanical movements of the GI tract.

Click on *Exercise 5.8* within your online platform or enter the address below into your web browser.
http://learn.genetics.utah.edu/content/metabolism/digestion/

Please read the information provided on the website.

EXERCISE 5.9
SWALLOWING

Purpose of exercise: To demonstrate the mechanical movements of the GI tract.

Click on *Exercise 5.9* within your online platform or enter the address below into your web browser.
https://medlineplus.gov/ency/anatomyvideos/000126.htm

Please read the overview and watch the video on this website. Use the space below to write a brief summary of what you learned.

_____

_____

EXERCISE 5.10
DENTAL SURGERIES

Purpose of exercise: To explore various types of surgeries associated with the teeth.

***Please make sure that your Adobe Flash Player is updated on your computer. Also, please be care not to click on any of the third-party advertisements because they will route you to another site.

You will perform the following virtual surgeries: **Dental Braces, Teeth Whitening, Teeth Cleaning, Root Canal Procedure, Dental Crown Placement, Wisdom Tooth Extraction,** and **Dental Filling.**

Enter the following addresses for each surgery type into your web browser. <u>Once you are ready to begin the surgery, click **START**. Please listen and read the instructions provided on the website and follow the interactive steps to perform the various surgeries.</u>

Click on *Exercise 5.10 (a)* within your online platform or enter the address below into your web browser.
Dental Braces: http://www.surgerysquad.com/surgeries/braces/

Click on *Exercise 5.10 (b)* within your online platform or enter the address below into your web browser.
Teeth Whitening: http://www.surgerysquad.com/surgeries/teeth-whitening/

Click on *Exercise 5.10 (c)* within your online platform or enter the address below into your web browser.
Teeth Cleaning: http://www.surgerysquad.com/surgeries/dental-prophylaxis-teeth-cleaning/

Click on *Exercise 5.10 (d)* within your online platform or enter the address below into your web browser.
Root Canal Procedure: http://www.surgerysquad.com/surgeries/virtual-root-canal/

Click on *Exercise 5.10 (e)* within your online platform or enter the address below into your web browser.
Dental Crown Placement: http://www.surgerysquad.com/surgeries/virtual-dental-crown/

Click on *Exercise 5.10 (f)* within your online platform or enter the address below into your web browser.
Wisdom Tooth Extraction: http://www.surgerysquad.com/surgeries/virtual-wisdom-tooth-extraction/

Click on *Exercise 5.10 (g)* within your online platform or enter the address below into your web browser.
Dental Filling: http://www.surgerysquad.com/surgeries/virtual-dental-filling-composite-amalgam/

EXERCISE 5.11
PERISTALSIS

Purpose of exercise: To demonstrate the mechanical movements of the GI tract.

Click on *Exercise 5.11* within your online platform or enter the address below into your web browser.
https://medlineplus.gov/ency/anatomyvideos/000097.htm

Please read the overview and watch the video on this website. Use the space below to write a brief summary of what you learned.

_____
_____
_____
_____
_____
_____
_____
_____
_____
_____
_____
_____
_____
_____
_____
_____

EXERCISE 5.12
DIGESTION GAME

Purpose of exercise: To identify the organs of the gastrointestinal tract and the accessory organs of digestion and their functions in digestion.

Click on *Exercise 5.12* within your online platform or enter the address below into your web browser.
http://interactivehuman.blogspot.com/2008/05/digestion-interactive-game-for-kids.html

Please read the information provided on the website.

EXERCISE 5.13
STOMACH

Purpose of exercise: To identify the organs of the gastrointestinal tract and the accessory organs of digestion and their functions in digestion.

Click on *Exercise 5.13* within your online platform or enter the address below into your web browser.
http://www.wesnorman.com/stomach.htm

Please read the information provided on the website.

EXERCISE 5.14
STOMACH ULCER

Purpose of exercise: To provide students with an understanding of gastrointestinal system disorders.

Click on *Exercise 5.14* within your online platform or enter the address below into your web browser.
https://medlineplus.gov/ency/anatomyvideos/000133.htm

Please read the overview and watch the video on this website. Use the space below to write a brief summary of what you learned.

_____
_____
_____
_____
_____
_____
_____
_____
_____
_____
_____
_____
_____
_____

EXERCISE 5.15
STOMACH, SPLEEN AND LIVER DISSECTION

Purpose of exercise: To describe the anatomy of the stomach, spleen, and liver through dissection.

Click on *Exercise 5.15* within your online platform or enter the address below into your web browser.
http://act.downstate.edu/courseware/haonline/labs/L38/ld0000.htm

Please read the information provided on the website. Click on each step to view the dissection.

EXERCISE 5.16
HEART BURN

Purpose of exercise: To provide students with an understanding of gastrointestinal system disorders.

Click on *Exercise 5.16* within your online platform or enter the address below into your web browser.
https://medlineplus.gov/ency/anatomyvideos/000068.htm

Please read the overview and watch the video on this website. Use the space below to write a brief summary of what you learned.

_____
_____
_____
_____
_____
_____
_____
_____
_____
_____
_____
_____
_____
_____

EXERCISE 5.17
INTESTINES AND PANCREAS DISSECTION

Purpose of exercise: To describe the anatomy of the intestines and pancreas through dissection.

Click on *Exercise 5.17* within your online platform or enter the address below into your web browser.
http://act.downstate.edu/courseware/haonline/labs/L39/ld0000.htm

Please read the information provided on the website. Click on each step to view the dissection.

EXERCISE 5.18
DUODENUM

Purpose of exercise: To identify the organs of the gastrointestinal tract and the accessory organs of digestion and their functions in digestion.

Click on *Exercise 5.18* within your online platform or enter the address below into your web browser.
http://www.wesnorman.com/duodenum.htm

Please read the information provided on the website.

EXERCISE 5.19
JEJUNUM AND ILEUM

Purpose of exercise: To identify the organs of the gastrointestinal tract and the accessory organs of digestion and their functions in digestion

Click on *Exercise 5.19* within your online platform or enter the address below into your web browser.
http://www.wesnorman.com/jejunumileum.htm

Please read the information provided on the website.

EXERCISE 5.20
LARGE INTESTINE

Purpose of exercise: To identify the organs of the gastrointestinal tract and the accessory organs of digestion and their functions in digestion

Click on *Exercise 5.20* within your online platform or enter the address below into your web browser.
http://www.wesnorman.com/largeintestine.htm

Please read the information provided on the website.

EXERCISE 5.21
RADIOLOGICAL INTERPRETATION – ABDOMINAL X-RAY

Purpose of exercise: To provide students a foundational knowledge in the art of radiological interpretation while reinforcing anatomy and physiology concepts.

During a radiographic procedure, an x-ray beam is passed through the body. A portion of the x-rays are absorbed or scattered by the internal structure and the remaining x-ray pattern is transmitted to a detector so that an image may be recorded for later evaluation. The recoding of the pattern may occur on film or through electronic means. X-rays are a type of radiation called electromagnetic waves. X-ray images show the parts of your body in different shades of black (radiolucent) and white (radiopaque). This is because different tissues absorb different amounts of radiation. Calcium in bones absorbs x-rays the most, so bones look white. Fat and other soft tissues absorb less, and look gray. Air absorbs the least, so lungs look black.

The most familiar use of x-rays is checking for broken bones, but x-rays are also used in other ways. For example, chest x-rays can spot pneumonia. Mammograms use x-rays to look for breast cancer. It is used to diagnose or treat patients by recording images of the internal structure of the body to assess the presence or absence of disease, foreign objects, and structural damage or anomaly.

This website allows students to build their skills in medical x-ray imaging by giving free access to materials. **YOU ARE NOT REQUIRED TO BUY ANYTHING ON THE RADIOLOGYMASTER CLASS WEBSITE FOR THIS COURSE**.

Click on *Exercise 5.21(a)* within your online platform or enter the address below into your web browser before starting this exercise: https://www.radiologymasterclass.co.uk/tutorials/tutorials_howto

Now click on *Exercise 5.21(b)* within your online platform or enter the address below into your web browser:
https://www.radiologymasterclass.co.uk/tutorials/abdo/abdomen_x-ray/anatomy_introduction

Abdominal X-ray
A systematic approach to abdominal X-ray interpretation is therefore relatively straightforward. This involves assessment of the bowel gas pattern, soft tissue structures, and bones. Full assessment includes a check of patient data, image quality, and checking for artifact and abnormal calcification. Please read and follow the instructions provided on the website. Click the "**Next >>**" button to move through the images on the page.

EXERCISE 5.22
COLORECTAL CANCER RISK ASSESSMENT TOOL

Purpose of exercise: To provide students with an understanding of gastrointestinal system disorders.

Click on *Exercise 5.22* within your online platform or enter the address below into your web browser.
https://www.cancer.gov/colorectalcancerrisk/tool.aspx

Please read the information provided on the website.

EXERCISE 5.23
LIVER

Purpose of exercise: To identify the organs of the gastrointestinal tract and the accessory organs of digestion and their functions in digestion

Click on *Exercise 5.23* within your online platform or enter the address below into your web browser.
http://www.wesnorman.com/liver.htm

Please read the information provided on the website.

EXERCISE 5.24
HEPATITIS RISK ASSESSMENT

Purpose of exercise: To provide students with an understanding of gastrointestinal system disorders.

Click on *Exercise 5.24* within your online platform or enter the address below into your web browser.
https://www2a.cdc.gov/hepatitis/RiskAssessment/start.html

Please read the information provided on the website.

EXERCISE 5.25
ORDERING LABS - COMPREHENSIVE METABOLIC PANEL (CMP)

Purpose of exercise: To provide students with knowledge about common laboratory tests ordered by healthcare professionals to help diagnose gastrointestinal medical conditions.

Click on *Exercise 5.25* within your online platform or enter the address below into your web browser.
https://labtestsonline.org/understanding/analytes/cmp/tab/test/

Answer the questions about the lab test below using the hyperlink/website listed in this exercise. Summarize your answer so that is fits within the space provided.

1. Formal name:
2. Also known as:
3. How is it used?

4. When is it ordered?

5. What does the test result mean?

6. How is the sample collected for testing?

EXERCISE 5.26
ORDERING LABS - ALANINE AMINOTRANSFERASE (ALT)

Purpose of exercise: To provide students with knowledge about common laboratory tests ordered by healthcare professionals to help diagnose gastrointestinal medical conditions.

Click on *Exercise 5.26* within your online platform or enter the address below into your web browser.
https://labtestsonline.org/understanding/analytes/alt/tab/test/

Answer the questions about the lab test below using the hyperlink/website listed in this exercise. Summarize your answer so that is fits within the space provided.

1. Formal name:
2. Also known as:
3. How is it used?

4. When is it ordered?

5. What does the test result mean?

6. How is the sample collected for testing?

EXERCISE 5.27
ALCOHOL AND YOU: AN INTERACTIVE BODY

Purpose of exercise: To demonstrate metabolic rate and the role of the liver in metabolism.

Click on *Exercise 5.27* within your online platform or enter the address below into your web browser.
https://www.collegedrinkingprevention.gov/SpecialFeatures/interactiveBody.aspx

Please read and follow the instructions provided on the website.

EXERCISE 5.28
ALCOHOL CALORIE CALCULATOR

Purpose of exercise: To demonstrate metabolic rate and the role of the liver in metabolism.

Click on *Exercise 5.28* within your online platform or enter the address below into your web browser.
https://www.collegedrinkingprevention.gov/specialfeatures/calculators/alcoholcaloriecalculator.aspx

Please read and follow the instructions provided on the website.

EXERCISE 5.29
PANCREAS

Purpose of exercise: To identify the organs of the gastrointestinal tract and the accessory organs of digestion and their functions in digestion

Click on *Exercise 5.29* within your online platform or enter the address below into your web browser.
http://www.wesnorman.com/pancreas.htm

Please read the information provided on the website.

EXERCISE 5.30
CALCULATE BODY MASS INDEX

Purpose of exercise: To describe the process of absorption of fats, carbohydrates, and proteins.

Click on *Exercise 5.30* within your online platform or enter the address below into your web browser.
https://www.nhlbi.nih.gov/health/educational/lose_wt/BMI/bmicalc.htm

Please read and follow the instructions provided on the website

EXERCISE 5.31
BODY WEIGHT PLANNER

Purpose of exercise: To describe the process of absorption of fats, carbohydrates, and proteins.

Click on *Exercise 5.31* within your online platform or enter the address below into your web browser.
https://www.supertracker.usda.gov/bwp/

Please read and follow the instructions provided on the website

EXERCISE 5.32
WEIGHT LOSS SURGERY- LAPAROSCOPIC GASTRIC BYPASS SURGERY

Purpose of exercise: To provide students with an understanding of gastrointestinal system disorders.

Click on *Exercise 5.32* within your online platform or enter the address below into your web browser.
https://www.broadcastmed.com/5005/videos/laparoscopic-gastric-bypass-surgery?view=displayPageNLM

Please watch the surgical video on this website. Use the space below to write a brief summary of what you learned.

_____
_____
_____
_____
_____
_____
_____
_____
_____
_____
_____
_____
_____
_____

EXERCISE 5.33
WEIGHT LOSS SURGERY- SLEEVE GASTRECTOMY

Purpose of exercise: To provide students with an understanding of gastrointestinal system disorders.

Click on *Exercise 5.33* within your online platform or enter the address below into your web browser.
https://www.broadcastmed.com/4120/videos/surgical-weight-loss-sleeve-gastrectomy?view=displayPageNLM

Please watch the surgical video on this website. Use the space below to write a brief summary of what you learned.

EXERCISE 5.34
ANAL PAIN CHECKUP

Purpose of exercise: To provide students with an understanding of gastrointestinal system disorders.

Click on *Exercise 5.34* within your online platform or enter the address below into your web browser.
http://www.freemd.com/anal-pain/visit-virtual-doctor.htm

Please read and follow the instructions provided on the website.

EXERCISE 5.35
RECTAL BLEEDING CHECKUP

Purpose of exercise: To provide students with an understanding of gastrointestinal system disorders.

Click on *Exercise 5.35* within your online platform or enter the address below into your web browser.
http://www.freemd.com/rectal-bleeding/overview.htm

Please read and follow the instructions provided on the website.

EXERCISE 5.36
ORDERING LABS – CALPROTECTIN

Purpose of exercise: To provide students with knowledge about common laboratory tests ordered by healthcare professionals to help diagnose gastrointestinal medical conditions.

Click on *Exercise 5.36* within your online platform or enter the address below into your web browser.
https://labtestsonline.org/understanding/analytes/calprotectin/tab/test/

Answer the questions about the lab test below using the hyperlink/website listed in this exercise. Summarize your answer so that is fits within the space provided.

1. Formal name:
2. Also known as:
3. How is it used?

4. When is it ordered?

5. What does the test result mean?

6. How is the sample collected for testing?

EXERCISE 5.37
GASTROINTESTINAL SURGERIES

Purpose of exercise: To explore types of breast examines and surgeries.

***Please make sure that your Adobe Flash Player is updated on your computer. Also, please be care not to click on any of the third-party advertisements because they will route you to another site.
You will perform the following virtual surgeries: **Colonoscopy, Laparoscopic, Appendectomy,** and **Gastric Bypass.**

Enter the following addresses for each surgery type into your web browser. <u>Once you are ready to begin the surgery, click **START**. Please listen and read the instructions provided on the website and follow the interactive steps to perform the various surgeries.</u>

Click on *Exercise 5.37 (a)* within your online platform or enter the address below into your web browser.
Colonoscopy: http://www.surgerysquad.com/surgeries/colonoscopy/

Click on *Exercise 5.37 (b)* within your online platform or enter the address below into your web browser.
Laparoscopic: http://www.surgerysquad.com/surgeries/laparoscopic-gallbladder-removal/

Click on *Exercise 5.37 (c)* within your online platform or enter the address below into your web browser.
Appendectomy: http://www.surgerysquad.com/surgeries/appendectomy-surgery/

Click on *Exercise 5.37 (d)* within your online platform or enter the address below into your web browser.
Gastric Bypass: http://www.surgerysquad.com/surgeries/rny-gastric-bypass-surgery/

EXERCISE 5.38
THE VIRTUAL AUTOPSY

Purpose of exercise: To teach students how to think critically about information they have learned and how to use it to answer questions about a person's illness or cause of death in case scenarios

Click on *Exercise 5.38* within your online platform or enter the address below into your web browser.
http://www.le.ac.uk/pa/teach/va/titlpag1.html

Please complete the following **Cases: 13 and 14.** Read and follow the instructions provided on the website.

EXERCISE 5.39
REVIEW QUESTIONS

Please continue by answering the questions below.

DIGESTIVE

1. The portion of the digestive system that extends from the mouth to the anus is the _____.
2. The liver and pancreas are part of the portion of the digestive system called the _____.
3. The stomach is located on the _____ side of the abdominal cavity just below the _____.
4. The pancreas is located between the _____ medially and the _____ laterally.
5. The changing of food to smaller pieces is called _____ digestion, and it creates more _____.
6. The end products of carbohydrate digestion are _____.
7. The end products of protein digestion are _____.
8. The _____ salivary glands are located below and in front of the ear.
9. The _____ salivary glands are located below the floor of the mouth.
10. The _____ of saliva dissolves food so that it can be tasted.
11. The _____ in saliva inhibits the growth of some bacteria.
12. The _____ in saliva digests starch to maltose.
13. The _____ of a tooth provides the chewing surface, and is made of _____.
14. The root of a tooth is made of _____, and is anchored in its socket by the _____ membrane.
15. The part of a tooth that contains blood vessels and nerves is the _____.

16. The cranial nerve pair that provides sensation for teeth is the _____.
17. The cranial nerve pair that provides sensation for teeth is the _____.
18. During swallowing, the opening from the pharynx to the larynx is covered by the _____.
19. During swallowing, the nasopharynx is covered by the _____.
20. The esophagus carries food from the _____ to the _____.
21. Backup of food from the stomach to the esophagus is prevented by the _____.
22. The lower esophageal sphincter prevents backup of food from the _____ to the _____.
23. The layer of the alimentary tube that produces digestive enzymes is the _____.
24. The layer of the alimentary tube that contains Meissner's plexus is the _____.
25. The layer of the alimentary tube that provides peristalsis is the _____.
26. The serous membrane that lines the abdominal cavity is the _____.
27. The external muscle layer of the alimentary tube is made of _____ tissue and is responsible for _____.
28. The cranial nerve pair that supplies the alimentary tube is the _____, and its effect on peristalsis is to _____ it.
29. Backup of chyme from the small intestine to the stomach is prevented by the _____.
30. The pyloric sphincter prevents backup of chyme from the _____ to the _____.
31. In the gastric mucosa, parietal cells secrete _____, chief cells secrete _____, and G cells secrete _____.
32. The parts of the small intestine, in order from the stomach, are the _____, the _____, and the _____.
33. Bile is produced by the _____ and stored in the _____.
34. The digestive function of bile is to _____.
35. The function of the gallbladder is to _____ and _____ bile.
36. Bile leaves the liver by way of the _____ duct and enters the gallbladder by way of the _____ duct.
37. The cystic duct is a two-way duct, carrying bile into or out of the _____.
38. Contraction of the gallbladder is stimulated by the hormone _____.
39. Secretion of pancreatic enzymes is stimulated by the hormone _____, and secretion of bicarbonates is stimulated by the hormone _____.
40. The function of bicarbonate pancreatic juice is to neutralize the _____ that comes from the _____.
41. Pancreatic amylase contributes to the digestion of _____.
42. Pancreatic trypsin contributes to the digestion of _____.
43. Pancreatic lipase contributes to the digestion of _____.
44. Protein is digested by pancreatic _____, starches by pancreatic _____, and fats by pancreatic _____.
45. The ileum of the _____ empties into the _____ of the colon.
46. The _____ prevents backup of feces from the colon to the small intestine.
47. The normal flora of the colon are the _____ that benefit us by producing _____.
48. Voluntary control of the defecation reflex is provided by the _____.
49. When blood glucose level is low, the liver changes its stored _____ to _____.
50. Chewing is an example of _____ digestion.

THE URINARY SYSTEM
LAB 6

CRASHCOURSE VIDEO(S):

Click on the video embedded within your online platform or enter the address below into your web browser:

1. **https://youtu.be/l128tW1H5a8**
2. **https://youtu.be/DlqyyyvTI3k**

(Please make sure to watch the video before continuing)

DEFINING KEY TERMS:

1. Countercurrent mechanism
2. Dialysis
3. Glomerular (Bowman's) capsule
4. Glomerular filtrate
5. Glomerular filtration
6. Glomerular Hydrostatic Pressure (GHSP)
7. Juxtaglomerular Apparatus (JGA)
8. Kidney
9. Metabolic waste
10. Renal clearance
11. Renal tubule
12. Renin-Angiotensin system
13. Tubular reabsorption
14. Tubular secretion
15. Ureters

16. Urethra

17. Urinalysis

18. Urinary bladder

19. Urinary incontinence

20. Urinary tract infection (UTI)

EXERCISE 6.1
OVERVIEW OF THE URINARY SYSTEM

Purpose of exercise: To provide a general overview of the urinary system.

Click on *Exercise 6.1* within your online platform or enter the address below into your web browser.
http://interactivehuman.blogspot.com/2008/06/animation-kidney-parts-of-nephron.html

Please read and watch the animation on the website.

EXERCISE 6.2
IDENTIFYING URINARY TISSUES UNDER THE MICROSCOPE

Purpose of exercise: To observe examples of the major urinary tissue types under the microscope

Click on *Exercise 6.2 (a)* within your online platform or enter the address below into your web browser:
http://histologyguide.org/

Once you enter the site, click onto the **Slide Box** tab located on the left hand side. You will see a list of slide categorized by tissue type and organ system in bold font. Click onto the tabs that correctly identifies the tissue type you must observe for this exercise. The tissue type is listed on the table below.

If you encounter difficulties linking to this web address, or would like to view a different source try the link below.

Click on *Exercise 6.2 (b)* within your online platform or enter the address below into your web browser:
http://www.kumc.edu/instruction/medicine/anatomy/histoweb/index.htm

(Your instructor may choose to upload pictures of the tissue types in your college's online platform. If this is the case, you can still utilize the listed slides below as a reference to identify.)

Make a drawing of each prepared slide on high power in the circles below.

Prepared Slide *(Tissue Example)*
1. Kidney
2. Ureter
3. Bladder
4. Transitional Epithelia

Observations

Tissue Type	Tissue Drawing on High Power (40X Objective)
1. Kidney	
2. Ureter	
3. Bladder	

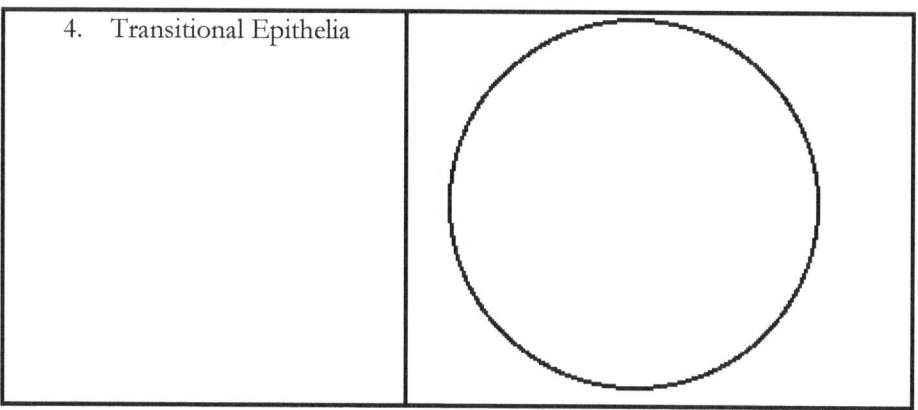

| 4. Transitional Epithelia | |

EXERCISE 6.3
ANATOMICAL PARTS

Purpose of exercise: To examine the structure associated with the urinary system.

Click on *Exercise 6.3* within your online platform or enter the address below into your web browser:
https://www.biodigital.com/

- *(This is a free site, but you will need to sign up with your name and a validated email address. You can use your personal email address or school email address. MAKE SURE TO CREATE A PASSWORD YOU CAN REMEMBER. I WOULD SUGGEST WRITING IT DOWN AND KEEPING IT IN A SAFE PLACE. YOU WILL USE IT AGAIN.)*

Click **SIGN UP**. Once you have provided the appropriate information, you are now able to access BioDigital's website content. Click **LOG IN** using the email and password you created. Then click **SIGN IN**. On the left of your screen choose from the systems listed.

Please click on the square boxes next to this icon. . Now click on the **Anatomy By Systems** icon. Click on the following image(s) with caption. **Male Urinary System, Female Urinary System.** View the structures associated with this system.

EXERCISE 6.4
BLADDER FUNCTION

Purpose of exercise: To demonstrate the structure and physiology of the urinary bladder.

Click on *Exercise 6.4* within your online platform or enter the address below into your web browser.
https://medlineplus.gov/ency/anatomyvideos/000009.htm

Please read the overview and watch the video on this website. Use the space below to write a brief summary of what you learned.

_____
_____
_____
_____
_____
_____
_____
_____
_____
_____
_____
_____
_____
_____

EXERCISE 6.5
UNDERSTANDING OF THE KIDNEYS

Purpose of exercise: To convey the role of the kidney.

Click on *Exercise 6.5* within your online platform or enter the address below into your web browser.
http://www.kidneypatientguide.org.uk/whattheydo.php

Please read the information provided on the website.

EXERCISE 6.6
STRUCTURE OF THE KIDNEY

Purpose of exercise: To identify the external and internal gross anatomical features of the kidneys.

Click on *Exercise 6.6* within your online platform or enter the address below into your web browser.
http://www.wesnorman.com/posteriorabdomen.htm

Please read the information provided on the website.

EXERCISE 6.7
LABELING THE KIDNEY

Purpose of exercise: To identify the external and internal gross anatomical features of the kidneys.

Label the parts of the kidney.

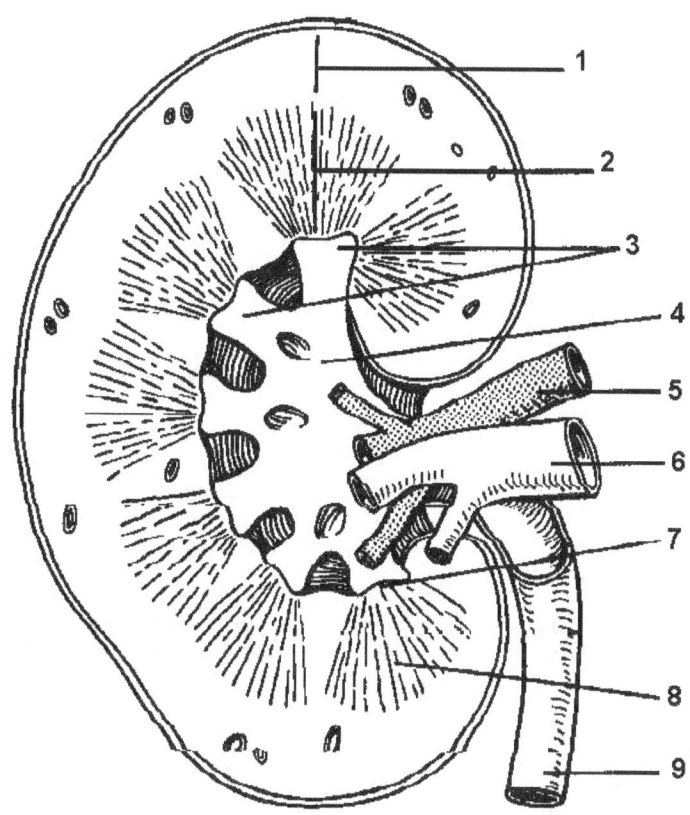

EXERCISE 6.8
FILTRATION OF THE KIDNEYS

Purpose of exercise: To discuss the formation of urine explaining the microscopic anatomy of the nephron and its basic functions of filtration, reabsorption, and secretion.

Our kidneys remove toxins from our blood: The renal artery brings blood into the kidneys which then process the blood, removing any unwanted substances and eliminating the waste in the urine. The kidneys then return the processed blood to the body through the renal vein. You will use everyday kitchen equipment to create an experiment that demonstrates the basic workings of the kidneys.

Materials
- 1/2 spoonful of crushed chalk
- 1/2 cup of water
- Two clear glass jars
- Red food coloring
- Coffee filter
- Rubber band

Procedure
1. Mix 1/2 spoonful of crushed chalk with 1/2 cup of water in a clear glass jar.
2. Add a few drops of red food coloring to the water. The chalk will represent toxins present in the blood, while the water will represent the blood.
3. Place a coffee filter over the top of the second jar and secure it with a rubber band. The filter will represent the kidneys that filter the toxins from the blood.
4. Pour the chalk/water mixture through the coffee filter into the second jar.
5. Observe how the filter traps the chalk as the colored water drips into the second jar. This helps illustrate how blood circulates through the kidneys, which then trap the toxins before returning the purified blood to the body's circulatory system.

EXERCISE 6.9
FLUID BALANCE

Purpose of exercise: To explain the role of key hormones on the kidney and their role in water and electrolyte balance.

Click on *Exercise 6.9* within your online platform or enter the address below into your web browser.
http://www.kidneypatientguide.org.uk/fluidAnim.php

Please read and watch the animation on the website.

EXERCISE 6.10
THE RENIN ANGIOTENSIN MECHANISM

Purpose of exercise: To discuss the role of the kidney in maintaining blood pressure and the function of the juxtaglomerular apparatus.

Click on *Exercise 6.10 (a)* within your online platform or enter the address below into your web browser.
http://mechanismsincardiology.com/free_animation.php?id=95

Please read and watch the animation on the website. Then move to the next list below.

Click on *Exercise 6.10(b)* within your online platform or enter the address below into your web browser.
http://mechanismsincardiology.com/free_animation.php?id=86

EXERCISE 6.11
KIDNEY QUIZ

Purpose of exercise: To provide a general overview of the urinary system.

Click on *Exercise 6.11* within your online platform or enter the address below into your web browser.
https://www.kidney.org/kidneydisease/kidneyQuiz

Please read and watch the four animation on the website.

EXERCISE 6.12
KIDNEY STONES

Purpose of exercise: To provide students with an understanding of urinary disorders.

Click on *Exercise 6.12* within your online platform or enter the address below into your web browser.
https://medlineplus.gov/ency/anatomyvideos/000075.htm

Please read the overview and watch the video on this website. Use the space below to write a brief summary of what you learned.

_____
_____
_____
_____
_____
_____
_____
_____
_____
_____
_____
_____
_____
_____

EXERCISE 6.13
URINALYSIS

Purpose of exercise: To provide students with knowledge about common laboratory tests ordered by healthcare professionals to help diagnose urinary medical conditions.

Click on *Exercise 6.13 (a)* within your online platform or enter the address below into your web browser.
http://www.nlm.nih.gov/medlineplus/ency/article/003579.htm

Use the following website to take notes:

1. What does a urinalysis typically test for in the urine?

2. What are two methods for collecting urine?

3. **Urinalysis checks for:**
   a. Physical color appearance

   b. Microscopic appearance

   c. Chemical appearance

4. What are four major reasons a urinalysis may be done?
   a.

   b.

   c.

   d.

5. What are considered to be normal urine test results?

6. What does abnormal urine test results mean?

Click on *Exercise 6.13 (b)* within your online platform or enter the address below into your web browser.

http://www.mayoclinic.org/tests-procedures/urinalysis/details/results/rsc-20255397

Explain indicators of disease that are present in urine. What does their presence in urine mean for the patient?

Acidity (pH)	
Specific gravity	
Protein	
Sugar	
Ketones	
Bilirubin	
Nitrites	
Blood	
Leukocytes	
Erythrocytes	
Bacteria or Yeasts	
Casts	
Crystals	

EXERCISE 6.14
HEMODIALYSIS

Purpose of exercise: To identify medical procedures used to treat urinary disorders.

Click on *Exercise 6.14* within your online platform or enter the address below into your web browser.
http://www.kidneypatientguide.org.uk/HDanim.php

Please read and watch the animation on the website.

EXERCISE 6.15
PERITONEAL DIALYSIS

Purpose of exercise: To identify medical procedures used to treat urinary disorders.

Click on *Exercise 6.15* within your online platform or enter the address below into your web browser.
http://www.kidneypatientguide.org.uk/pdanim.php

Please read and watch the animation on the website.

EXERCISE 6.16
KIDNEY TRANSPLANT

Purpose of exercise: To identify medical procedures used to treat urinary disorders.

Click on *Exercise 6.16* within your online platform or enter the address below into your web browser.
http://www.kidneypatientguide.org.uk/TRAanim.php

Please read and watch the animation on the website.

EXERCISE 6.17
KIDNEY TRANSPLANTATION

Purpose of exercise: To identify medical procedures used to treat urinary disorders.

Click on *Exercise 6.17* within your online platform or enter the address below into your web browser.
http://nlm.bcst.md/videos/living-donor-kidney-transplant?view=displayPageNLM

Please watch the surgical video on this website. Use the space below to write a brief summary of what you learned.

_____

_____

EXERCISE 6.18
KIDNEY TRANSPLANTATION

Purpose of exercise: To identify medical procedures used to treat urinary disorders.

Click on *Exercise 6.18* within your online platform or enter the address below into your web browser.
https://www.kidney.org/atoz/content/kidney-transplant

Answer the questions below using the information from the website.

1. What is a kidney transplant?

2. What is a "preemptive" or "early" transplant?

3. Who can get a kidney transplant?

4. What if I'm older or have other health problems?

5. How will I pay for a transplant?

6. How do I start the process of getting a kidney transplant?

7. How does the evaluation process work?

8. What does the operation involve?

9. What are anti-rejection medicines?

10. What happens after I go home?

11. What if my body tries to reject the new kidney?

12. How often do rejection episodes happen?

13. When can I return to work?

14. Will I need to follow a special diet?

15. Where do donated kidneys come from?

16. Are there disadvantages to living donation?

17. What are the financial costs to the living donor?

EXERCISE 6.19
URINARY INCONTINENCE – MALE OR FEMALE CHECKUP

Purpose of exercise: To provide students with an understanding of urinary disorders.

Click on *Exercise 6.19* within your online platform or enter the address below into your web browser.
http://www.freemd.com/urinary-incontinence/overview.htm

Please read and follow the instructions provided on the website.

EXERCISE 6.20
ADVANCED PROCEDURES IN MALE INCONTINENCE: THE VIRTUE® MALE SLING

Purpose of exercise: To identify medical procedures used to treat urinary disorders.

Click on *Exercise 6.20* within your online platform or enter the address below into your web browser.
https://www.broadcastmed.com/3922/videos/advanced-procedures-in-male-incontinence-the-virtue-male-sling?view=displaypageNLM

Please watch the surgical video on this website. Use the space below to write a brief summary of what you learned.

EXERCISE 6.21
ORDERING LABS - URINE CULTURE

Purpose of exercise: To provide students with knowledge about common laboratory tests ordered by healthcare professionals to help diagnose urinary conditions.

Click on *Exercise 6.21* within your online platform or enter the address below into your web browser.
https://labtestsonline.org/understanding/analytes/urine-culture/tab/test/

Answer the questions about the lab test below using the hyperlink/website listed in this exercise. Summarize your answer so that is fits within the space provided.

1. Formal name:
2. Also known as:
3. How is it used?

4. When is it ordered?

5. What does the test result mean?

6. How is the sample collected for testing?

EXERCISE 6.22
URINARY TRACT INFECTIONS

Purpose of exercise: To provide students with an understanding of urinary disorders.

Click on *Exercise 6.22 (a)* within your online platform or enter the address below into your web browser.
http://medicalcenter.osu.edu/patientcare/healthcare_services/womens_health/urinary/uti/Pages/index.aspx

1. What causes urinary tract infections?

2. Which bacteria most likely lead to a UTI?

3. Where does this bacteria normally live?

Click on *Exercise 6.22 (b)* within your online platform or enter the address below into your web browser.
http://medicalcenter.osu.edu/patientcare/healthcare_services/womens_health/urinary/uti/Pages/index.aspx

4. Describe the different types of urinary tract infections:
   a. Urethritis

   b. Cystitis

   c. Pyelonephritis

5. What are the symptoms of a urinary tract infection?

Click on *Exercise 6.22 (c)* within your online platform or enter the address below into your web browser.
http://medicalcenter.osu.edu/patientcare/healthcare_services/womens_health/urinary/uti/Pages/index.aspx

6. Describe the four ways urinary tract infections are diagnosed:
   a. Urinalysis

   b. Intravenous Pyelogram (IVP)

   c. Cystoscopy (Also called cystourethroscopy)

   d. Renal ultrasound

7. What are some treatments for urinary tract infections?

8. Name five steps that can be taken to prevent urinary tract infections:
   a.
   b.
   c.
   d.
   e.

EXERCISE 6.23
URINARY TRACT INFECTION (UTI) CHECKUP

Purpose of exercise: To provide students with an understanding of urinary disorders.

Click on *Exercise 6.23* within your online platform or enter the address below into your web browser.
http://www.freemd.com/urinary-tract-infection/overview.htm

Please read and follow the instructions provided on the website.

EXERCISE 6.24

THE VIRTUAL AUTOPSY

Purpose of exercise: To teach students how to think critically about information they have learned and how to use it to answer questions about a person's illness or cause of death in case scenarios

Click on *Exercise 6.24* within your online platform or enter the address below into your web browser.
http://www.le.ac.uk/pa/teach/va/titlpag1.html

Please complete the following **Cases: 15 and 18.** Read and follow the instructions provided on the website.

EXERCISE 6.25
REVIEW QUESTIONS

Please continue by answering the questions below.

URINARY

1. The kidneys are located _____ to the spine, _____ to the diaphragm, and _____ to the peritoneum.
2. The kidneys are retroperitoneal, which means they are _____.
3. The kidneys are cushioned by _____ and are held in place by the _____.
4. A ureter extends from the _____ to the _____.
5. The male urethra is enclosed by the _____ and _____.
6. The outer layer of kidney tissue is called the _____.
7. The inner layer of kidney tissue is called the _____.
8. The term renal pyramids is another name for the _____.
9. The cavity on the medial side of a kidney is called the _____.
10. The renal cortex is made of _____ and _____ (parts of a nephron).
11. In the kidney, loops of Henle and collecting tubules are located in the _____.
12. In the kidney, calyces are part of the _____.
13. In the kidney, the renal pelvis has funnel-shaped extensions called _____.
14. In the kidney, the calyces are part of the _____ and enclose the tips of the _____.
15. The function of the nephrons of a kidney is to form _____ from _____.
16. The two major parts of a nephron are the _____ and _____.
17. A renal corpuscle consists of a(n) _____ and _____.
18. A glomerulus is a network of _____, and is surrounded by a _____.
19. A Bowman's capsule has _____ layers, and the layer that is very permeable is the _____ layer.
20. The energy for renal filtration is provided by _____.
21. Movement of substances from the glomerulus to Bowman's capsule takes place by the process of _____.
22. The process of glomerular filtration is selective in terms of the _____ of the materials in the blood.
23. In the kidney, the renal tubules are surrounded by the _____.

24. In the kidney, tubular reabsorption takes place from the _____ to the _____.
25. In tubular reabsorption, glucose is reabsorbed by the process of _____.
26. In tubular reabsorption, positive ions are reabsorbed by the process of _____.
27. In tubular reabsorption, water is reabsorbed by the process of _____.
28. In the proximal convoluted tubule, the surface area of the cells is increased by the presence of _____.
29. The maximum amount of glucose that can be reabsorbed by the kidney tubules is called the _____.
30. The process of tubular secretion may include the nitrogenous waste product _____.
31. The renal artery is a branch of the _____.
32. The renal vein carries blood to the _____.
33. In the glomerulus, blood pressure is high because the _____ arteriole has a smaller diameter than the _____ arteriole.
34. In the kidneys, the effect of ADH is to _____.
35. In the kidneys, the hormone that increases the reabsorption of sodium ions and water is _____.
36. In the kidneys, the hormone that increases the excretion of sodium ions and water is _____.
37. In the kidneys, the hormone that increases the excretion of potassium ions is _____.
38. In the kidneys, the hormone that increases the reabsorption of calcium ions is _____.
39. In the kidneys, the hormone that increases the reabsorption of sodium ions is _____, and the hormone that increases the secretion of sodium ions is _____.
40. If ADH secretion decreases, more _____ will be excreted in urine.
41. If aldosterone secretion decreases, more _____ ions will be present in urine and more _____ ions will remain in the blood.
42. The kidneys respond to increased acidity of the blood by excreting more _____ ions in order to _____ the pH of the blood.
43. The kidneys respond to increased alkalinity of the blood by retaining more _____ ions in order to _____ the pH of the blood.
44. When the kidneys retain hydrogen ions to regulate pH, the ions that will be excreted in urine are _____ and _____.
45. The purpose of the renin-angiotensin mechanism is to _____.
46. The juxtaglomerular cells of the kidney secrete _____ when _____ decreases.
47. When blood pressure decreases, the kidneys secrete the enzyme _____, which will start the process of the formation of _____.
48. The functions of angiotensin II are to cause _____ and to increase the secretion of _____.
49. The kidneys secrete erythropoietin during a state of _____.
50. The function of erythropoietin is to stimulate the _____ to increase the rate of _____.
51. The part of the urinary bladder that expels urine is the _____.
52. The _____ is usually contracted to prevent outflow of urine.
53. The trigone is on the _____ of the urinary bladder, and its boundaries are the openings of the _____ and _____.
54. The normal pH range of urine is _____ to _____.
55. The concentrating ability of the kidneys is reflected in the measurement of the _____ of urine.
56. The nitrogenous waste product that comes from the use of excess amino acids for energy production is _____.
57. The nitrogenous waste product that comes from energy production in muscles is _____.
58. The nitrogenous waste product that comes from the metabolism of nucleic acids is _____.
59. If the kidneys are not functioning properly, blood levels of the nitrogenous waste products will _____, and urine levels will _____.

60. If the kidneys are not functioning properly, the levels of the nitrogenous waste products will decrease in the _____ and increase in the _____.

THE REPRODUCTIVE SYSTEM
LAB 7

CRASHCOURSE VIDEO(S):

Click on the video embedded within your online platform or enter the address below into your web browser:

1. https://youtu.be/RFDatCchpus
2. https://youtu.be/-XQcnO4iX_U
3. https://youtu.be/SUdAEGXLO-8
4. https://youtu.be/BtsSbZ85yiQ

(Please make sure to watch the video before continuing)

DEFINING KEY TERMS:

1. Acrosome:

2. Androgens:

3. Breast Cancer:

4. Cervix:

5. Cesarean:

6. Chromosomes:

7. Circumcision:

8. Diploid:

9. Ejaculation:

10. Emission:

11. Endometriosis:

12. Endometrium:

13. Epididymis:

14. Erection:

15. Estrogen:

16. Fallopian tubes:

17. Fetus:

18. Flagellum:

19. Gametes:

20. Haploid:

21. Hysterectomy:

22. Mammary glands:

23. Mastectomy:

24. Maturation:

25. Meiosis:

26. Menopause:

27. Myometrium:

28. Oocyte:

29. Oogenesis:

30. Orgasm:

31. Oviduct:

32. Ovulation:

33. Perimetrium:

34. Perineum:

35. Placenta:

36. Prepuce:

37. Prostate gland:

38. Scrotum:

39. Seminal vesicles:

40. Sperm:

41. Spermatogenesis:

42. Testes:

43. Testosterone:

44. Varicocele:

45. Vas deferens:

46. Vasectomy:

47. Venereal disease:

48. Zygote:

EXERCISE 7.1
REPRODUCTIVE OVERVIEW

Purpose of exercise: To provide a general overview of the reproductive system.

Click on *Exercise 7.1* within your online platform or enter the address below into your web browser.
http://www.wesnorman.com/thoraxlesson2.htm

Please read the information provided on the website.

EXERCISE 7.2
ANATOMICAL PARTS

Purpose of exercise: To examine the structure associated with the reproductive system.

Click on *Exercise 7.2* within your online platform or enter the address below into your web browser:
https://www.biodigital.com/

- *(This is a free site, but you will need to sign up with your name and a validated email address. You can use your personal email address or school email address. MAKE SURE TO CREATE A PASSWORD YOU CAN REMEMBER. I WOULD SUGGEST WRITING IT DOWN AND KEEPING IT IN A SAFE PLACE. YOU WILL USE IT AGAIN.)*

Click **SIGN UP**. Once you have provided the appropriate information, you are now able to access BioDigital's website content. Click **LOG IN** using the email and password you created. Then click **SIGN IN**. On the left of your screen choose from the systems listed.

Please click on the square boxes next to this icon. . Now click on the **Anatomy By Systems** icon. Click on the following image(s) with caption. **Male Reproductive System, Female Reproductive System.** View the structures associated with this system.

EXERCISE 7.3
IDENTIFYING MALE REPRODUCTIVE TISSUE UNDER THE MICROSCOPE

Purpose of exercise: To observe examples of the major male reproductive tissue types under the microscope.

Click on *Exercise 7.3 (a)* within your online platform or enter the address below into your web browser:
http://histologyguide.org/

Once you enter the site, click onto the **Slide Box** tab located on the left hand side. You will see a list of slide categorized by tissue type and organ system in bold font. Click onto the tabs that correctly identifies the tissue type you must observe for this exercise. The tissue type is listed on the table below.

If you encounter difficulties linking to this web address, or would like to view a different source try the link below.

Click on *Exercise 7.3 (b)* within your online platform or enter the address below into your web browser:
http://www.kumc.edu/instruction/medicine/anatomy/histoweb/index.htm

(Your instructor may choose to upload pictures of the tissue types in your college's online platform. If this is the case, you can still utilize the listed slides below as a reference to identify.)

Make a drawing of each prepared slide on high power in the circles below.

Prepared Slide *(Tissue Example)*
1. Testis
2. Epididymis
3. Ductus Deferens
4. Seminal Vesicle
5. Prostate
6. Penis

Observations

Tissue Type	Tissue Drawing on High Power (40X Objective)
1. Testis	
2. Epididymis	
3. Ductus Deferens	
4. Seminal Vesicle	

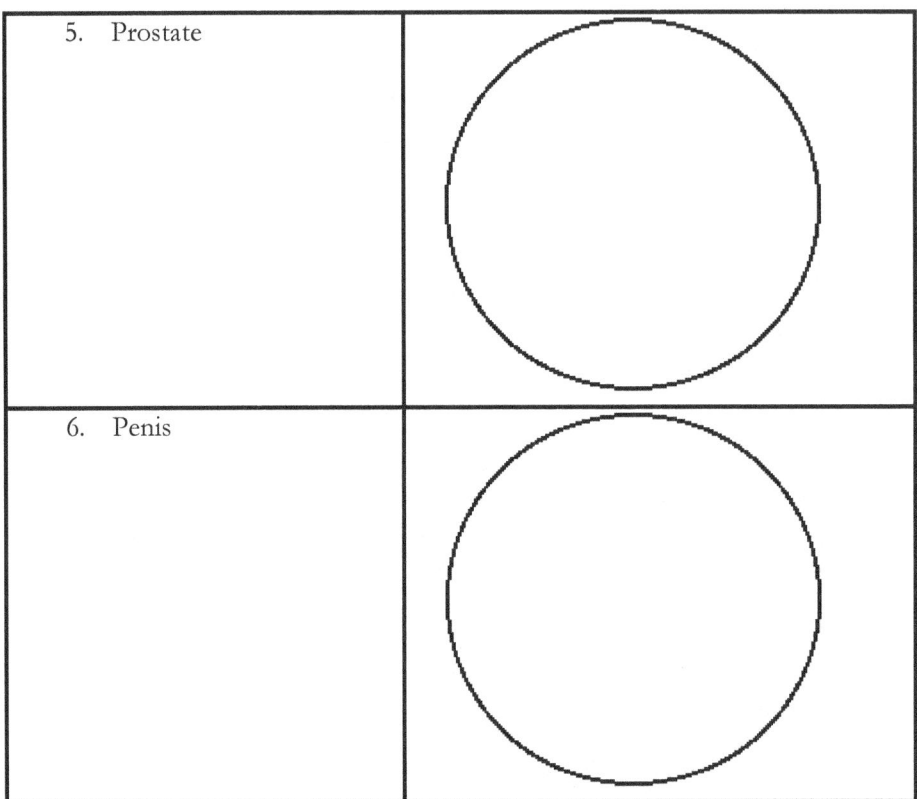

EXERCISE 7.4
THE MALE PERINEUM AND THE PENIS DISSECTION

Purpose of exercise: To describe the anatomy of the male perineum and the penis through dissection.

Click on *Exercise 7.4* within your online platform or enter the address below into your web browser.
http://act.downstate.edu/courseware/haonline/labs/L42/lo0000.htm

Please read the information provided on the website. Click on each step to view the dissection.

EXERCISE 7.5
INGUINAL REGION, SCROTUM AND TESTES DISSECTION

Purpose of exercise: To describe the anatomy of the male inguinal region, scrotum and testes through dissection

Click on *Exercise 7.5* within your online platform or enter the address below into your web browser.
http://act.downstate.edu/courseware/haonline/labs/L36/ld0000.htm

Please read the information provided on the website. Click on each step to view the dissection.

EXERCISE 7.6
SPERM RELEASE PATHWAY

Purpose of exercise: To explain the structure and functions of the male reproductive organs and the pathway of sperm.

Click on *Exercise 7.6* within your online platform or enter the address below into your web browser.
https://medlineplus.gov/ency/anatomyvideos/000121.htm

Please read the overview and watch the video on this website. Use the space below to write a brief summary of what you learned.

_____
_____
_____
_____
_____
_____
_____
_____
_____
_____
_____
_____
_____
_____
_____
_____
_____

EXERCISE 7.7
VASECTOMY

Purpose of exercise: To identify medical procedures used for male sterilization.

Click on *Exercise 7.7* within your online platform or enter the address below into your web browser.
https://medlineplus.gov/ency/anatomyvideos/000139.htm

Please read the overview and watch the video on this website. Use the space below to write a brief summary of what you learned.

EXERCISE 7.8
BREAST LUMP CHECKUP

Purpose of exercise: To provide students with an understanding of reproductive disorders.

Click on *Exercise 7.8* within your online platform or enter the address below into your web browser.
http://www.freemd.com/breast-lump/overview.htm

Please read and follow the instructions provided on the website.

EXERCISE 7.9
PATERNITY TEST

Purpose of exercise: To identify medical procedures used for paternity testing.

Click on *Exercise 7.9* within your online platform or enter the address below into your web browser.
http://www.sumanasinc.com/webcontent/animations/content/paternitytesting.html

Please read and watch the animation on the website. Follow the instructions on the page.

EXERCISE 7.10
MALE CIRCUMCISION

Purpose of exercise: To identify reproductive medical procedures.

Click on *Exercise 7.10* within your online platform or enter the address below into your web browser.
https://www.broadcastmed.com/4229/videos/circumcision?view=displayPageNLM

Please watch the surgical video on this website. Use the space below to write a brief summary of what you learned.

_____
_____
_____
_____
_____
_____
_____
_____
_____
_____
_____
_____
_____
_____
_____
_____
_____

EXERCISE 7.11
TESTICULAR DISORDERS - VARICOCELE SURGERY

Purpose of exercise: To provide students with an understanding of reproductive disorders.

Click on *Exercise 7.11* within your online platform or enter the address below into your web browser.
http://nlm.bcst.md/videos/varicocele-surgery?view=displayPageNLM

Please watch the surgical video on this website. Use the space below to write a brief summary of what you learned.

_____
_____
_____
_____
_____

EXERCISE 7.12
IDENTIFYING FEMALE REPRODUCTIVE TISSUE UNDER THE MICROSCOPE

Purpose of exercise: To observe examples of the major female reproductive tissue types under the microscope.

Click on *Exercise 7.12 (a)* within your online platform or enter the address below into your web browser:
http://histologyguide.org/

Once you enter the site, click onto the **Slide Box** tab located on the left hand side. You will see a list of slide categorized by tissue type and organ system in bold font. Click onto the tabs that correctly identifies the tissue type you must observe for this exercise. The tissue type is listed on the table below.

If you encounter difficulties linking to this web address, or would like to view a different source try the link below.

Click on *Exercise 7.12 (b)* within your online platform or enter the address below into your web browser:
http://www.kumc.edu/instruction/medicine/anatomy/histoweb/index.htm

(Your instructor may choose to upload pictures of the tissue types in your college's online platform. If this is the case, you can still utilize the listed slides below as a reference to identify.)

Make a drawing of each prepared slide on high power in the circles below.

Prepared Slide *(Tissue Example)*
1. Ovary
2. Oviduct
3. Uterus
4. Cervix
5. Vagina
6. Mammary Gland
7. Placenta
8. Umbilical Cord

Observations

Tissue Type	Tissue Drawing on High Power (40X Objective)
1. Ovary	
2. Oviduct	
3. Uterus	
4. Cervix	

5. Vagina	
6. Mammary Gland	
7. Placenta	
8. Umbilical Cord	

EXERCISE 7.13
THE FEMALE PERINEUM DISSECTION

Purpose of exercise: To describe the anatomy of the female perineum through dissection.

Click on *Exercise 7.13* within your online platform or enter the address below into your web browser.
http://act.downstate.edu/courseware/haonline/labs/L41/ld0000.htm

Please read the information provided on the website. Click on each step to view the dissection.

EXERCISE 7.14
OVULATION

Purpose of exercise: To demonstrate female fertility.

Click on *Exercise 7.14* within your online platform or enter the address below into your web browser.
https://medlineplus.gov/ency/anatomyvideos/000094.htm

Please read the overview and watch the video on this website. Use the space below to write a brief summary of what you learned.

_____
_____
_____
_____
_____
_____
_____
_____
_____
_____
_____
_____
_____
_____
_____

EXERCISE 7.15
PREGNANCY TEST

Purpose of exercise: To demonstrate medical tests used to determine pregnancy.

Click on *Exercise 7.15* within your online platform or enter the address below into your web browser.
http://www.sumanasinc.com/webcontent/animations/content/pregtest.html

Please read and watch the animation on the website.

EXERCISE 7.16
HOW SEX IS DETERMINED

Purpose of exercise: To provide an understanding about how gender is determined.

Click on *Exercise 7.16* within your online platform or enter the address below into your web browser.
http://www.pbs.org/wgbh/nova/miracle/dete_flash.html

Please read and watch the animation on the website.

EXERCISE 7.17
GENDER TESTING OF ATHLETES

Purpose of exercise: To provide an understanding about how gender is determined.

Click on *Exercise 7.17* within your online platform or enter the address below into your web browser.
http://www.hhmi.org/biointeractive/gender-testing-athletes

Please read the information provided on the website. Click **"START CLICK AND LEARN"** to complete the interactive exercise.

EXERCISE 7.18
MASTECTOMY

Purpose of exercise: To identify medical procedures used to treat female reproductive disorders.

Click on *Exercise 7.18* within your online platform or enter the address below into your web browser.
http://nlm.bcst.md/videos/breast-reconstruction-deep-inferior-epigastric-perforator?view=displayPageNLM

Please watch the surgical video on this website. Use the space below to write a brief summary of what you learned.

_____
_____
_____

EXERCISE 7.19
BREAST AWARENESS & SURGERIES FOR CANCER

Purpose of exercise: To explore types of breast examines and surgeries.

***Please make sure that your Adobe Flash Player is updated on your computer. Also, please be care not to click on any of the third-party advertisements because they will route you to another site.

You will perform the following virtual surgeries: **Breast Cancer Awareness, Lumpectomy, Mastectomy,** and **Double Mastectomy.**

Enter the following addresses for each surgery type into your web browser. <u>Once you are ready to begin the surgery, click **START**. Please listen and read the instructions provided on the website and follow the interactive steps to perform the various surgeries.</u>

Click on *Exercise 7.19 (a)* within your online platform or enter the address below into your web browser.
Breast Cancer Awareness: http://www.surgerysquad.com/surgeries/breast-cancer-overview/

Click on *Exercise 7.19 (b)* within your online platform or enter the address below into your web browser.
Lumpectomy: http://www.surgerysquad.com/surgeries/virtual-lumpectomy-surgery/

Click on *Exercise 7.19 (c)* within your online platform or enter the address below into your web browser.
Mastectomy: http://www.surgerysquad.com/surgeries/virtual-mastectomy/

Click on *Exercise 7.19 (d)* within your online platform or enter the address below into your web browser.
Double Mastectomy: http://www.surgerysquad.com/surgeries/double-mastectomy-surgery/

EXERCISE 7.20
COSMETIC BREAST SURGERIES

Purpose of exercise: To explore two types of cosmetic breast surgeries.

***Please make sure that your Adobe Flash Player is updated on your computer. Also, please be care not to click on any of the third-party advertisements because they will route you to another site.

You will perform the following virtual surgeries: **Silicone Breast Implants** and **Saline Breast Implants.**

Enter the following addresses for each surgery type into your web browser. Once you are ready to begin the surgery, click **START**. Please listen and read the instructions provided on the website and follow the interactive steps to perform the various surgeries.

Click on *Exercise 7.20 (a)* within your online platform or enter the address below into your web browser.
Silicone Breast Implants: http://www.surgerysquad.com/surgeries/silicone-breast-implants/

Click on *Exercise 7.20 (b)* within your online platform or enter the address below into your web browser.
Saline Breast Implants: http://www.surgerysquad.com/surgeries/saline-breast-implants/

EXERCISE 7.21
HYSTERECTOMY

Purpose of exercise: To identify medical procedures used to treat female reproductive disorders.

Click on *Exercise 7.21* within your online platform or enter the address below into your web browser.
https://www.broadcastmed.com/3925/videos/scarless-hysterectomy?view=displayPageNLM

Please watch the surgical video on this website. Use the space below to write a brief summary of what you learned.

_____
_____
_____
_____
_____
_____
_____
_____
_____
_____
_____
_____
_____

EXERCISE 7.22
WINDOWS OF THE WOMB

Purpose of exercise: To provide an understanding about human conception.

Click on *Exercise 7.22* within your online platform or enter the address below into your web browser.
http://www.pbs.org/wgbh/nova/miracle/wind_flash.html

Please read and watch the animation on the website.

EXERCISE 7.23
CESAREAN SECTION

Purpose of exercise: To identify medical procedures used in childbirth.

Click on *Exercise 7.23* within your online platform or enter the address below into your web browser.
https://medlineplus.gov/ency/anatomyvideos/000028.htm

Please read the overview and watch the video on this website. Use the space below to write a brief summary of what you learned.

_____
_____
_____
_____
_____
_____
_____
_____
_____
_____
_____
_____
_____
_____
_____

EXERCISE 7.24
GYNECOLOGICAL SURGERIES

Purpose of exercise: To explore various types of gynecological surgeries.

***Please make sure that your Adobe Flash Player is updated on your computer. Also, please be care not to click on any of the third-party advertisements because they will route you to another site.
You will perform the following virtual surgeries: **Natural Child Birth, Child Birth with Epidural,** and **C-Section.**

Enter the following addresses for each surgery type into your web browser. <u>Once you are ready to begin the surgery, click **START**. Please listen and read the instructions provided on the website and follow the interactive steps to perform the various surgeries.</u>

Click on *Exercise 7.24 (a)* within your online platform or enter the address below into your web browser.
Natural Child Birth: http://www.surgerysquad.com/surgeries/natural-child-birth/

Click on *Exercise 7.24 (b)* within your online platform or enter the address below into your web browser.
Child Birth with Epidural: http://www.surgerysquad.com/surgeries/child-birth-with-epidural/

Click on *Exercise 7.24 (c)* within your online platform or enter the address below into your web browser.
C-Section: http://www.surgerysquad.com/surgeries/virtual-c-section-cesarean-surgery/

EXERCISE 7.25
VENEREAL DISEASE – (STD) CHECKUP

Purpose of exercise: To provide students with an understanding of reproductive disorders.

Click on *Exercise 7.25* within your online platform or enter the address below into your web browser.
http://www.freemd.com/venereal-disease/overview.htm

Please read and follow the instructions provided on the website.

EXERCISE 7.26
DISCHARGE CHECKUP

Purpose of exercise: To provide students with an understanding of reproductive disorders.

Click on *Exercise 7.26 (a)* within your online platform or enter the address below into your web browser. IF YOU ARE A FEMALE. http://www.freemd.com/vaginal-discharge/overview.htm

Click on *Exercise 7.26 (b)* within your online platform or enter the address below into your web browser. IF YOU ARE A MALE. http://www.freemd.com/penile-discharge/overview.htm

Please read and follow the instructions provided on the website.

EXERCISE 7.27
THE VIRTUAL AUTOPSY

Purpose of exercise: To teach students how to think critically about information they have learned and how to use it to answer questions about a person's illness or cause of death in case scenarios

Click on *Exercise 7.27* within your online platform or enter the address below into your web browser.
http://www.le.ac.uk/pa/teach/va/titlpag1.html

Please complete the following **Case: 6**. Read and follow the instructions provided on the website.

EXERCISE 7.28
REVIEW QUESTIONS

Please continue by answering the questions below.

REPRODUCTION

1. Spermatogenesis takes place in the _____, and each cell that undergoes meiosis produces _____ functional sperm cell(s).
2. Oogenesis takes place in the _____, and each cell that undergoes meiosis produces _____ functional egg cell(s).
3. In a woman's lifetime, the process of oogenesis begins at _____ and ends at _____.
4. In the testes, the hormone that initiates sperm production is _____.
5. In the testes, the hormone that initiates sperm production is _____.
6. In the ovary, the hormone that initiates development of an ovum is _____.
7. In the ovary, the hormone that initiates development of an ovum is _____, which is secreted by the _____ gland.
8. The corpus luteum of the ovary secretes the hormones _____ and _____.
9. The hormone that stimulates secretion of progesterone is _____, which is secreted by the _____ gland.
10. The testes are located in the _____, because sperm production requires a lower _____.
11. The epididymis carries sperm from the _____ to the _____.
12. The _____ carries sperm from the testis to the ductus deferens.
13. The interstitial cells of the testes produce _____.
14. From the epididymis, the ducts that carry sperm are the _____, _____, and _____.
15. From the ductus deferens, the ducts that carry sperm are the _____ and the _____.
16. The male reproductive glands that produce secretions to promote sperm motility are the _____, _____, and _____.
17. The male reproductive glands that produce secretions to promote sperm motility are the _____, _____, and _____.
18. Semen has an _____ pH, which is important because the female vagina has an _____ pH.
19. A sperm cell contains _____ (number of) chromosomes, which are found in the _____.

20. The _____ of a sperm cell contains the chromosomes, the _____ provide ATP, and the _____ provides motility.
21. The ovaries are located in the _____ cavity, and are _____ to the uterus.
22. A developing ovarian follicle produces the hormone _____ when stimulated by _____.
23. In the ovary, a mature follicle is called a _____ follicle, and after it ruptures it becomes the _____.
24. A fallopian tube extends from the _____ laterally to the _____ medially.
25. The fimbriae of a fallopian tube enclose the _____.
26. The two tissues of a fallopian tube that propel the ovum (or zygote) toward the uterus are _____ and _____.
27. The function of the ciliated epithelium that lines a fallopian tube is to _____.
28. The uterus is _____ to the urinary bladder and _____ to the ovaries.
29. The exit of the uterus, for menstrual flow, is the _____.
30. The upper part of the uterus is the _____, and the _____ enter this part.
31. The layer of the uterine wall that contracts for delivery of an infant is the _____.
32. The myometrium of the uterus is made of _____ tissue, and its function is to _____.
33. The lining of the uterus is the _____.
34. The hormones that are directly necessary for the growth of blood vessels in the uterine lining are _____ and _____.
35. The part of the endometrium that is lost in menstruation is the _____.
36. The part of the endometrium that is not lost in menstruation is the _____.
37. The birth canal for the infant at the end of gestation is the _____.
38. The vagina is between the _____ anteriorly and the _____ posteriorly.
39. The urethral and vaginal openings are covered by the _____ and the _____.
40. The female external genital organs are collectively called the _____.
41. The female external genital structure that has an important sensory function is the _____.
42. The _____ secrete mucus at the vaginal orifice.
43. The perineum is the area between the _____ anteriorly and the _____ posteriorly.
44. The perineum refers to the _____ floor.
45. In the mammary glands, milk is produced by the _____.
46. In the mammary glands, the lactiferous ducts all converge at the _____.
47. The secretory portions of the mammary glands are surrounded and cushioned by _____.
48. Human milk contains the carbohydrate _____ as a nutrient.
49. The hormone that stimulates growth of the duct system of the mammary glands is _____.
50. The hormone that stimulates growth of the secretory cells of the mammary glands is _____.
51. In the mammary glands, the hormone _____ stimulates milk production, and the hormone _____ stimulates release of milk.
52. In the mammary glands, the hormone _____ stimulates growth of the duct system, and the hormone _____ stimulates growth of the secretory cells.
53. The menstrual cycle requires FSH and LH from the _____ gland, and estrogen and progesterone from the _____.
54. In the menstrual cycle, the ovum matures during the _____ phase.
55. In the menstrual cycle, ovulation is stimulated by the hormone _____.
56. In the menstrual cycle, the hormone _____ stimulates the ruptured ovarian follicle to become the corpus luteum.
57. In the menstrual cycle, the endometrium is shed during the _____ phase.

EXERCISE 7.29
VIRTUAL FETAL PIG DISSECTION

Purpose of exercise: To familiarize students with all of the significant anatomical structures of the human body through dissection using the fetal pig specimen as a model.

Click on *Exercise 7.29* within your online platform or enter the address below into your web browser.
https://www.whitman.edu/academics/departments-and-programs/biology/virtual-pig

Please read and follow the instructions provided on the website. Familiarize yourself with the virtual fetal pig dissection at this website. Use the checklist below to learn the different parts. Choose and view different parts of the Virtual Fetal Pig Dissection by clicking on the organ system tabs located on the left of the webpage. Place a check next to the anatomical part once you have identified it.

FETAL PIG LAB CHECK LIST

Circulatory System and Heart
- Atria ☐
- Ventricles ☐
- Pericardium ☐
- Coronary Arteries ☐
- Anterior and Posterior Vena Cava ☐
- Pulmonary Artery ☐
- Ductus Arteriosus ☐
- Aorta and Aortic Arch ☐
- Abdominal Aorta ☐
- Left and Right Subclavian ☐
- Bicarotid ☐ Left and Right Carotid ☐
- Intercostal arteries ☐
- Renal Artery ☐
- Mesenteric Artery ☐
- Hepatic Artery ☐
- Splenic Artery ☐
- External and Internal Iliac ☐
- Ilio-lumbar artery ☐
- Femoral & Deep Femoral Artery ☐
- Anterior & Posterior Tibial Artery ☐
- Umbilical Vessels ☐

Respiratory
- Diaphragm ☐
- Trachea ☐ Cartilaginous rings ☐
- Larynx ☐
- Epiglottis ☐
- Pharynx ☐
- Lungs ☐

Digestive System
- Tongue ☐
- Salivary Glands ☐
- Pharynx ☐
- Papillae (taste buds) ☐
- Soft palate ☐ Hard palate ☐
- Esophagus ☐

Liver ☐
Small Intestine ☐
Duodenum ☐
Ileum ☐
Large Intestine (colon) ☐
Cecum ☐
Gall bladder ☐
Bile duct ☐
Pancreas ☐
Pyloric Sphincter ☐
Cardiac Sphincter ☐
Mesentery ☐
Rectum ☐
Anus ☐

Urogenital (Reproductive, Excretory)
Kidneys ☐
Ureter ☐
Urinary bladder ☐
Urethra ☐
Urogenital opening ☐

Male Reproductive System
Scrotal sacs ☐
Testes ☐
Epididymis ☐
Vas deferens ☐
Penis ☐

Female Reproductive System
Ovaries ☐
Fallopian tubes ☐
Uterus ☐
Vagina ☐
Urogenital sinus ☐

Miscellaneous
Thyroid ☐
Spleen ☐
Masseter Muscle ☐
Lymph nodes ☐
Nipples ☐

www.ingramcontent.com/pod-product-compliance
Lightning Source LLC
Chambersburg PA
CBHW082328220526
45470CB00008B/2436